Studies in Systems, Decision and Control

Volume 583

The series "Studies in Systems, Decision and Control" (SSDC) covers both new developments and advances, as well as the state of the art, in the various areas of broadly perceived systems, decision making and control–quickly, up to date and with a high quality. The intent is to cover the theory, applications, and perspectives on the state of the art and future developments relevant to systems, decision making, control, complex processes and related areas, as embedded in the fields of engineering, computer science, physics, economics, social and life sciences, as well as the paradigms and methodologies behind them. The series contains monographs, textbooks, lecture notes and edited volumes in systems, decision making and control spanning the areas of Cyber-Physical Systems, Autonomous Systems, Sensor Networks, Control Systems, Energy Systems, Automotive Systems, Biological Systems, Vehicular Networking and Connected Vehicles, Aerospace Systems, Automation, Manufacturing, Smart Grids, Nonlinear Systems, Power Systems, Robotics, Social Systems, Economic Systems and other. Of particular value to both the contributors and the readership are the short publication timeframe and the world-wide distribution and exposure which enable both a wide and rapid dissemination of research output.

Indexed by SCOPUS, DBLP, WTI Frankfurt eG, zbMATH, SCImago.

All books published in the series are submitted for consideration in Web of Science.

Artur Zaporozhets · Mykhailo Kulyk ·
Vitalii Babak · Viktor Denysov

Structure Optimization of Power Systems with Renewable Energy Sources

Springer

Artur Zaporozhets
General Energy Institute
National Academy of Sciences of Ukraine
Kyiv, Ukraine

Mykhailo Kulyk
General Energy Institute
National Academy of Sciences of Ukraine
Kyiv, Ukraine

Vitalii Babak
General Energy Institute
National Academy of Sciences of Ukraine
Kyiv, Ukraine

Viktor Denysov
General Energy Institute
National Academy of Sciences of Ukraine
Kyiv, Ukraine

ISSN 2198-4182 ISSN 2198-4190 (electronic)
Studies in Systems, Decision and Control
ISBN 978-3-031-83696-1 ISBN 978-3-031-83697-8 (eBook)
https://doi.org/10.1007/978-3-031-83697-8

This Springer imprint is published by the registered company Springer Nature Switzerland AG
The registered company address is: Gewerbestrasse 11, 6330 Cham, Switzerland

Introduction

Relevance. The potential for sustainable economic development of the national economy is largely determined by the reliability and efficiency of energy supply systems. An important feature of the modern world is the openness of the national economy. At the same time, the functioning and development of the national economy take place under conditions of constantly increasing global competition and the influence of other economies, the specifics of the functioning of global resource markets, including fuel and energy, capital, as well as the development goals and actual actions of transnational corporations. Alongside the growing volatility of the global economic system, there is an increasing demand for the environmental friendliness of technological processes, particularly those involved in the extraction, conversion, and consumption of energy resources and carriers. Currently, there are no universal methods for economically acceptable and coherent solutions to the problem of meeting humanity's energy needs while ensuring no negative impact on the environment.

An important task is the improvement and development of new mathematical programming models for investigating the directions and optimal parameters of the technological renewal of energy system elements operating within energy associations. Incorporating economic and technological indicators of national economy and production development into these models, represented as quasi-dynamic functions with discrete stochastic variables, allows for the exploration of optimal volumes of technological renewal for integrated energy systems.

The development of theoretical foundations, methods, and tools for evaluating the efficiency of the functioning and development of energy systems, as well as mathematical methods and models for determining the possibilities and limitations of renewing elements of both autonomous and integrated energy systems, has been the focus of work by such Ukrainian and foreign scientists as M. M. Kulyk, V. P. Babak, O. V. Kyrylenko, A. F. Zharkin, A. V. Nosovsky, B. S. Stogniy, S. O. Kudrya, Yu. I. Yakymenko, I. P. Kondratenko, V. H. Kuznetsov, V. V. Mokhor, O. V. Novoseltsev, V. F. Reztsov, Yu. H. Stoyan, A. Lovins, M. Jacobson, D. Kammen, J. Goldemberg, V. Smil, D. MacKay, and others.

The relevance of this work is determined by current trends and challenges in the field of energy and sustainable development, considering the following key points:

1. Transition to Renewable Energy: Optimizing the structure of integrated energy systems with a focus on renewable energy sources and distributed generation is crucial for achieving sustainable development goals and mitigating climate change.
2. Technological Progress: Continuous advancements in generation technologies, energy storage, and smart grid development provide new opportunities for the optimal integration of renewable energy into existing and new energy systems. Models and tools that can promptly track the variability and intermittency of renewable sources are essential for making optimal management decisions to ensure the reliability and resilience of energy systems.
3. Environmental Impact: Optimizing energy systems using renewable sources helps reduce greenhouse gas emissions and minimize negative environmental impacts.
4. Integration of Energy Storage: Effective use of energy storage technologies is a key aspect of optimizing energy systems with a high share of renewable energy sources. Models that address the optimal integration of energy storage (batteries) and demand response strategies are crucial for maintaining the stability of electricity generation and supply.
5. Social Significance: Research into models and means of optimizing the implementation of renewable energy can contribute to creating effective scenarios for sustainably providing consumers with environmentally clean and economically affordable energy.
6. Economic Competitiveness: Reducing the costs of renewable energy technologies combined with the potential savings from optimizing the operation modes of traditional generation makes the integration of renewable energy economically attractive. Models that consider the economic aspects of integrating renewable energy are important for making strategic decisions in the energy sector.

Therefore, an important task of this research is the optimization of the structure and operating modes of energy systems in the context of the rapid growth of renewable energy sources (RES), which requires the development of new and improvement of existing methods, economic-mathematical models, and software tools.

Incorporating economic and technological indicators of national economic development into these models, represented as quasi-dynamic functions with discrete stochastic variables, allows for the exploration of optimal volumes of technological renewal for integrated energy systems.

Thus, the task of developing new and improving existing methods and economic-mathematical models for studying the directions and optimal parameters of the technological development of energy system elements is a relevant scientific problem that needs to be addressed.

Aim and Objectives of the Research. The aim of the research is to develop new and improve existing methods, economic-mathematical models, and software tools to optimize the structure and operating modes of generating, backup, and storage capacities of energy systems under the conditions of rapid growth of renewable energy capacities worldwide and in Ukraine.

In line with the defined aim, the research formulates and addresses the following objectives:

- Investigate the possibilities of optimized dispatching of energy system components, the permissible parameter ranges of generating, backup, and storage capacities that ensure stable and balanced operation of the integrated energy system.
- Improve economic-mathematical models for determining the technical and economic efficiency of the functioning of individual electricity generation facilities throughout their life cycle.
- Develop a method for economic-technological forecasting of the optimal functioning and development parameters of the energy system.
- Develop new and improve existing economic-mathematical programming models for investigating directions and determining optimal parameters for the technological development of energy system elements.
- Conduct computational experiments using the developed method, models, and software tools to identify short- and long-term development prospects for the Ukrainian energy system.

Research Methods. The methodological foundation of the conducted scientific research comprises the following methods:

- Mathematical Statistics: Analysis and forecasting of the behavior of stochastic time series.
- Mathematical Modeling Methods: Long-term energy forecasts, one of the main goals of which is to identify the main objective trends in the development of energy supply for the national economy.
- Modeling the Diffusion of Innovative Technologies in Energy.
- System Studies: In energy, a key aspect of which is mathematical modeling for optimal energy system management.
- Dynamic Computer Methods for Modeling Energy Systems: As hierarchical managed quasi-dynamic systems.

New Results.

1. *Development of a New Economic-Technological Forecasting Method*: A new method for economic-technological forecasting of the optimal functioning and development parameters of the energy system has been developed. This method, by using a modified integral quality criterion of the economic-technological impact functional, allows for the consideration of national economic and production development indicators in the form of quasi-dynamic functions with discrete stochastic variables. This is crucial for assessing the energy system's ability to meet energy and power consumption volumes and modes.
2. *Development of a New Mathematical Model for the Diffusion of New Technologies in Energy*: A new mathematical model for the diffusion of new technologies

in the energy sector has been developed. This model generalizes Frank Bass's mixed influence model for the diffusion of new technologies and, unlike existing models, includes the economic-technological impact functional. The developed model accounts for the stochasticity of economic and technological development parameters of the national economy when forecasting the development of the integrated energy system's structure amid the rapid growth of renewable energy capacities.

3. *Proposal of a Set of Regression Models*: The use of a set of regression models in the form of generalized logistic curves has been proposed. This approach improves forecasting efficiency, ensuring that the parameter values obtained by the least squares method do not contradict experimental data.
4. *Development of a New Long-Term Technological Renewal Model*: A new model for the long-term technological renewal of the integrated energy system structure has been developed. This model allows for the assessment of the potential for improving energy system components, considering the impact of economic-technological development indicators of the national economy and production in the form of quasi-dynamic functions with discrete stochastic variables.
5. *Generalization of the Classical Dynamic System Management Model*: The generalization of the classical dynamic system management model has led to the development of an economic-mathematical model of the energy system as a complex hierarchical quasi-dynamic system with levels of administrative-territorial hierarchy and sectoral (sub-sectoral) infrastructure, detailed by the structure of its technological content. Unlike known models, the developed model includes the economic-technological impact functional, which comprises elements of state matrices, control actions, and random elements of the external impact matrix, and also considers the peculiarities of the operation of generating, backup, and storage capacities of the energy systems of Ukraine and neighboring ENTSO-E member countries. This allowed for a significant refinement of the modeling results and the forecasting of the permissible limits of the most significant system parameters.

Practical Significance of the Obtained Results. Based on the proposed method of economic-technological forecasting of optimal functioning and development parameters of the energy system and economic-mathematical models, the author developed a problem-oriented software-information complex. This complex includes software modules for models optimizing the operating modes of energy systems. The practical results of the work were obtained through mathematical modeling using the developed software-information complex, specifically:

- Synchronization Conditions: Conditions for synchronizing the operation of Ukraine's unified energy system (UES) with neighboring countries that are part of ENTSO-E (Hungary, Slovakia, Poland, Romania) were determined, considering the limiting values of cross-border electricity exchange volumes, geographical location, availability, and capacity of interstate power lines, as well as the maximum permissible volumes of import and export capacity.

- Future Trends: It was shown that by 2040, the main trend in changes in the UES of Ukraine should be:
    (a) The increase and maximum utilization of nuclear power plant (NPP) capacity.
    (b) The increase in the capacity of interstate power lines, which is beneficial for both Ukraine and neighboring ENTSO-E countries, as it ensures the balance reliability of each country under conditions of increasing NPP and renewable energy sources (RES) capacities.
- Renewable Energy Integration: Despite technical progress in renewable energy, particularly the reduced cost of electricity production by wind and photovoltaic power plants, their integration into energy associations will only be feasible with the availability of sufficiently powerful, energy-efficient, and economically viable industrial energy storage systems due to the variability of electrical generation.
- Optimization of Generation and Storage Modes: A solution was proposed for forecasting the optimal modes of using the generating and storage capacities of the energy system based on the criterion of minimizing the weighted average cost of electricity production from traditional sources, RES, and energy storage technologies.

The results of the research have been implemented in a problem-oriented software-information complex, which includes software modules for the numerical realization of the developed and studied mathematical models. This complex is also included in the normative educational component "Mathematical Modeling and Decision Making in Energy Supply Systems" of the educational-professional master's program "Energy Management and Energy Efficient Technologies" under the specialty 141—Electrical Power Engineering, Electrical Engineering, and Electromechanics.

The use of the mathematical models and software tools developed in this work allows for the analysis and forecasting of current trends in the modernization of existing and the implementation of prospective energy generation technologies. It also helps in identifying the main directions and reference scenarios in this field.

Kyiv, Ukraine
August 2024

Artur Zaporozhets
Mykhailo Kulyk
Vitalii Babak
Viktor Denysov

Contents

Abbreviations

CBF	Cross-Border Flows
CCUS	Carbon Capture, Utilization, and Storage
CHP	Combined Heat and Power Plant
CIS	Commonwealth of Independent States
CUF	Capacity Utilization Factor
ECL	Electric Load Schedule
ENTSO-E	European Network of Transmission System Operators for Electricity
ESS	Energy Storage Systems
EU	European Union
GDP	Gross Domestic Product
HPP	Hydroelectric Power Plant
IEA	International Energy Agency
IPS	Integrated Power System
IRENA	International Renewable Energy Agency
LACE	Levelized Avoided Cost of Electricity
LCOE	Levelized Cost of Electricity
MHD Generators	Magnetohydrodynamic Generators
mini-HPP	Small Power Hydroelectric Plant
NPP	Nuclear Power Plant
PL	Power Line
PSPP	Pumped Storage Power Plant
RES	Renewable Energy Sources
SHW	Solid Household Waste
SEL	Schedule of Electrical Loads
SIC	Software and Information Complex
SPP	Solar Power Plant
TPP	Thermal Power Plant

TSOs	Transmission System Operators
UES	United Energy System
USA	United States of America
WPP	Wind Power Plant

Chapter 1
Current Trends in Modeling and Updating the Dissemination Processes of Energy Conversion and Utilization Technologies in the Energy Sector of Ukraine

1.1 System Innovations in Integrated Power Systems

Ensuring sustainable development in the energy sector is one of the main tasks of global economic development [1]. It is achieved through improving energy efficiency associated with the use of economically viable and environmentally "clean" energy resources. Authors identify the following new requirements for the energy sector: ensuring energy stability in crisis conditions, creating new energy generation, storage, and transmission technologies, transitioning to "smart" energy management, and developing autonomous energy sources. These new requirements are largely related to the implementation of the Smart Grid concept.

As indicated by the authors [2]: "At the beginning of the twenty-first century, there is a fundamental reorientation of energy policy in energy-dependent countries in connection with the transition from the "industrial" phase of economic and social development to the "post-industrial," "informational," and "knowledge societies." Energy policy is subordinated to environmental and social requirements, economic and organizational stability, and security. The key requirements for the new energy sector of the twenty-first century have become accessibility, reliability, cost-effectiveness, efficiency, harmony with the environment, and safety. The world has seen significant changes in energy development strategies that meet the requirements for sustainable development. The development of computer technologies and internet capabilities, the latest advancements in information and network technologies, information-control systems based on microprocessor and power electronics, along with the development of market relations in the energy industry, have become the basis for a qualitatively new approach to energy efficiency, leading to the emergence of a new type of energy—intelligent energy. A wide-ranging development of the "smart efficiency" concept has been highlighted, reflecting the intellectual interaction of valuation, production processes, and effective use of resources, embodied in the "Smart Grid" concept [2]. Authors [2] emphasize that three main characteristics can be distinguished in the "electric world": system intelligence, energy

A. Zaporozhets et al., *Structure Optimization of Power Systems with Renewable Energy Sources*, Studies in Systems, Decision and Control 583,
https://doi.org/10.1007/978-3-031-83697-8_1

security, and systematics—comprehensive regulation technologies, energy accumulation technologies, energy decentralization, new generation energy systems, and environmental sustainability—renewable energy sources. The concept of innovative transformation of the power industry implies a fully integrated, self-regulating, and self-renewing system with a network topology that includes all generating sources, main and distribution networks, and all types of electricity consumers interconnected by a unified network of real-time automated devices. It is evident that modern power engineering requires the formation of a comprehensive hierarchical control system with increasing automation capabilities and reliability across all electricity market participants: generators, networks, utility companies, and consumers.

System innovations that arise from the interaction of multiple components of the energy system are called systemic innovations [3]. They don't just embody technologies but encompass a broader context. They are essential for successfully transforming the energy sector and, ultimately, for energy transition. Systemic innovations play a key role in any transformation. They provide a high level of flexibility in the energy system for integrating a large share of variable renewable generation and can also facilitate smart electrification of various end-use sectors. Systemic innovations entail progress across all sectors, from technology providers and entrepreneurs to regulators, system operators, and even consumers. The focus is on having the necessary technologies, establishing the required regulatory framework for market participants and accounting for innovations. Additionally, systemic innovations enable innovative business models to benefit from smart electrification strategies.

Smart energy [4] can be considered one of the priority areas for implementing systemic innovations in Ukraine's energy sector. Analyzing the main global and domestic trends in smart energy development helps identify the primary focus for implementing smart energy in Ukraine. This focus is on establishing a unified state energy policy aimed at the global innovative model of restructuring the electricity market. This goal can be achieved by implementing modernized energy equipment and innovative technologies based on advanced "smart" energy grids (Smart Grid) in conditions of limited resources.

The trends in the development of innovative processes in Ukraine and their impact on the economy in modern conditions are discussed in [5]. According to the author, "The mechanism for making economic decisions regarding the development of an innovation strategy should be aimed at ensuring rapid modernization of the economy through extensive use of the latest domestic scientific and technical achievements."

According to the authors [6], the implementation of innovations in clean energy technologies plays a crucial role in accelerating the global energy transition. The goal of these innovations is to bring about positive changes in energy systems, one of which is the reduction of CO_2 emissions. Currently, the main focus is on collecting data on input parameters for innovative processes. There have been significantly fewer efforts to actively identify key performance indicators for tracking the results of innovations in clean energy technologies. This is due to the uncertainty and time lag between knowledge generation, codification, cost reduction, and increased utilization. However, such indicators allow for a more meticulous comparative analysis of the relative performance of innovation support for different technologies. Therefore, a significant portion of the mentioned research is dedicated to studying datasets, analyzing costs, and the efficiency of energy technologies.

The authors [7] highlight the use of experience from leading companies and advanced countries to enhance energy efficiency through the implementation of energy management systems. To address this issue from a management perspective, it is necessary to consistently apply a systemic approach to energy management. This approach can provide additional opportunities to improve energy efficiency. Systematic utilization of energy management allows for optimizing energy resource expenditures for any type and level of organization. To achieve this, the management cycle should be oriented towards updating and improving key energy characteristics.

The traditional energy sector, based on fossil fuels (coal, gas, oil, petroleum products), nuclear energy, hydroenergy, and renewable energy sources, are the four directions in energy highlighted by the authors [8]. Unconventional energy includes four main directions: renewable energy sources (solar energy, geothermal, wind, biomass, low-grade Earth heat, water, air, hydraulic, including mini-hydroelectric power stations, waves, tides, seawater ebb and flow); secondary renewable energy sources (heat from industrial and construction waste, heat and gas from waste incineration, solid household waste (SHW)); unconventional technologies for using non-renewable and renewable energy sources (hydrogen energy, gasification and pyrolysis, catalytic methods of combustion and processing of organic fuels, synthetic fuels); energy installations (or converters), such as heat pumps, hydrosteam turbines, direct energy conversion installations—fuel cells, photovoltaic converters, thermoelectric generators, thermoemission generators, MHD generators. There is a huge potential for energy conservation and an even greater potential for the implementation of renewable energy sources.

The model of an integrated complex energy supply system for consumers, in the form of an energy hub, is proposed in [9]. Environmental and economic factors are taken into account based on a multi-criteria approach. It is presented as a model for optimizing the operation modes of the integrated energy system. The model is designed to improve the efficiency of energy nodes where distributed generation sources operate. The objective function is formulated as follows:

$$\sum_{k=0}^{T} c(k)P(k) \rightarrow \min,$$

where $c(k)$ represents the tariffs for electricity, thermal energy, and gas on the market for the kth time interval (hour), in UAH/kW·h; $P(k)$ represents the volume of purchase of electricity, thermal energy, and gas on the market for the kth time interval (hour), in kW h. The balance of energy resource inflows and consumption is specified in the form of columns and constraint matrices.

The monograph [10] presents new developments in methods for determining energy efficiency indicators, provides methodological approaches to determining structural and technological potentials for energy conservation at different hierarchical levels, taking into account the influence of environmental factors. Improved methods for forecasting energy resource consumption are proposed, including a comprehensive method for harmonizing forecast parameters obtained at different levels of hierarchy, enhancing the accuracy of forecasts, and reducing the number of forecast variants. Forecast estimates of fuel consumption (in general, coal, natural gas, petroleum products), electricity, and thermal energy consumption are provided for the horizon up to 2040.

The authors [11] have examined a multi-factor model for medium-term and long-term forecasting of electricity consumption. They have presented calculations regarding the determination of monthly electricity consumption dependencies.

1.2 Potential and Directions for Improving Energy Efficiency in the World and in Ukraine

Energy transitions are ongoing processes worldwide. Despite sustainable solutions being projected for the future, many societies still operate with high-carbon energy and environmental pollution due to fossil fuels. Understanding the potential pathways for energy system transitions requires the integration of new energy technologies, environmental sciences, economics, and governance [12]. This work presents the results of studies on sustainable energy systems, innovative pathways, and management models of energy systems. Four trends in the transition to sustainable energy have been identified:

1. Economy and Management of Sustainable Energy.
2. Production and Consumption of Renewable Energy.
3. Environmental Impact of Energy Systems.
4. Electric Vehicles and Energy Storage.

Theoretical, technological, and innovative models, as well as successful examples, are discussed. It is emphasized that updated strategies and management models are necessary to stimulate the sustainable energy transition across various scales and sectors.

Despite progress in policies and measures to improve energy efficiency in recent years, programs that fully utilize Ukraine's potential in this area have not yet been established. An effective and balanced framework is needed, focusing on market prices, regulatory and control mechanisms, fiscal measures and tax incentives, and technology development [13].

The monograph [14] is dedicated to the conversion of various types of renewable energy sources into electrical and thermal energy. It examines efficient methods, tools, and the specific features of their use. Contemporary indicators of the energy potential and distribution of renewable energy sources across Ukraine are provided. Measures to improve the efficiency of renewable energy sources and storage systems have been proposed.

At the end of 2019, the International Energy Agency presented the "World Energy Outlook" [15] and the "World Energy Model" [16], which simulates the behavior of global energy markets. The following set of three scenarios is considered for this purpose:

1. Current Policies Scenario;
2. Stated Policies Scenario;
3. Sustainable Development Scenario.

The Stated Policies Scenario, previously known as the New Policies Scenario, forecasts an annual 1% increase in electricity demand until 2040. Renewable sources, mainly solar photovoltaics, account for two-thirds of this growth, with natural gas providing the remaining third. This leads to a reduction in demand for coal and oil. Consequently, there is an increase in the required energy storage capacity. All of this slows down emissions growth. However, this slowdown does not align with sustainable development goals.

The Sustainable Development Scenario defines the pathway to achieving energy goals through large-scale changes across all sectors of energy production and consumption. It fully aligns with Sustainable Development Goals related to energy access, air purification, and climate response. It includes limitations on global temperature rise to 1.5 °C by the end of the twenty-first century, which is lower than the 2 °C limit set by the Paris Agreement. This scenario involves increasing nuclear power capacities and significantly reducing coal usage. At the same time, gas generation capacities play a crucial role in providing flexibility and should be supported by CCUS (Carbon Capture, Utilization, and Storage) systems.

The common features of the stated policies and sustainable development scenarios, thanks to extensive innovation, are:

(1) The main energy carrier becomes electrical energy, ensuring energy and environmental security.
(2) An increase in energy efficiency and the efficient use of materials is forecasted.
(3) Strengthening of requirements for reducing emissions is increased, through:
(a) Gradual increase in CO_2 prices, reaching 100–140 US dollars per ton of CO_2 for the sustainable development scenario;
(b) Stricter emissions standards for SO_2 and NO for industrial facilities using solid fuels;
(c) Limiting the maximum sulfur content in oil products to 1% for heavy fuel oil and 10 mg/kg for gasoline and diesel.

The conclusions drawn from the research of the International Renewable Energy Agency (IRENA) in "Global Energy Transformation: Roadmap to 2050" [17] are similar:

- Expanded use of renewable energy sources.
- Strengthening electrification, with its share projected to increase from 20% in 2020 to 50% by 2050.

Therefore, energy efficiency can become a crucial factor in achieving key sustainable development goals in climate change. It is anticipated that the energy transition will also reduce costs and bring significant socio-economic benefits.

1.3 The Need for Improving Forecasting and Modeling Methods for the Operation and Recovery Modes of Energy Systems

Progress in digital and industrial technologies [18] has a significant impact on all sectors of the economy, including energy. Serious challenges in recent years have highlighted the need to increase the stability and flexibility of both technological development and the improvement of forecasting and modeling methods for the operation and recovery modes of energy systems. Considering the global trend towards digitalization, it is necessary to support modeling technological innovations in key sectors of industrial ecosystems while maintaining their ability to develop safely and steadily. The European Commission, in its document "Updating the New Industrial Strategy 2020: Building a Stronger Single Market for Europe's Recovery" [19], has supported the European Green Deal and the Circular Economy Action Plan, as well as digital strategies such as "Shaping Europe's Digital Future", "Data", "White Paper on Artificial Intelligence", and the "Digital Decade of Communications".

Energy systems [20] have a complex structure and contain a large number of elements. This leads to difficulties in modeling, researching, and managing the modes of their components. To overcome these difficulties, complex multifunctional software is required, which should be used to solve the tasks of assessing the current state and forecasting optimal operating modes of energy units. To address these tasks, the authors [20] proposed a multi-agent approach.

The correct approach to determining the energy efficiency of complex energy systems is quite problematic and requires the application of specialized methods [21]. According to the authors [21], energy efficiency can be understood as the most complete correspondence of the technical–economic essence of the system to consumer requirements. The organization of energy efficiency indicators can be carried out either on a "bottom-up" or "top-down" principle with gradual aggregation or disaggregation of macroeconomic indicators defined at each level of the system hierarchy.

The economic-mathematical model of production type proposed in the paper [22] allows for forecasting balanced supply of carbon-containing fuel, electricity, and heat from fossil fuels and renewable energy sources. According to the authors, the model is built based on open data sources and, unlike known models, uses subsystems for the production of abstract products. Selective aggregation procedures are also used to construct various forms of interproduct balance matrices.

The integer linear programming model for scheduling generating capacities to simultaneously cover electricity and heat load profiles is proposed in [23]. This model represents further development of existing versions of optimal generating capacity scheduling models for covering load profiles in power systems. The development is achieved by introducing additional equations that allow for the investigation of heat production technologies designed to cover thermal load profiles in centralized heating systems. This enables comprehensive research into the simultaneous operation of electricity and heat production technologies to meet the demand of local energy systems, addressing the pressing issue of distributed generation structure formation.

Integrated energy systems with multiple energy carriers [24] provide high efficiency in supplying electricity to consumers. The ability to transform energy systems under the influence of internal and external factors promotes the expansion of areas of their technological integration and wider use of integrated multi-carrier energy systems. For modeling such systems, the authors [24] proposed a concept and developed an imitational model of a power unit. The implementation of this imitational model was done using Matlab/Simulink. An example of the application of the imitational model is provided. The modeling results demonstrate the capabilities and prospects of using the imitational model of a power unit.

In the work [25], the issue of energy management (ILES) is considered, taking into account the energy balance and technological constraints for the operation of each power unit aimed at achieving economic efficiency of the system. A mathematical model for solving the energy management problem is presented, which allows optimizing operational costs.

The relevance of modeling energy systems is becoming increasingly important in the process of planning the implementation of the "energy transition," as emphasized in the work [26]. It is necessary to consider both cogeneration issues and problems of regional and temporal coordination. While many modern reviews group models by type or energy carrier, the authors propose a classification based on expert assessment quality. The advantages of dynamic cogeneration models of heat and electrical energy in combination with traditional and prospective control schemes are analyzed.

In the work [27], the authors presented a recently developed model called MESS—Multi Energy System Simulator, along with their assessment of its advantages and disadvantages compared to the optimization model Calliope [28]. They believe that using their simplified method of modeling energy systems on an urban scale yields satisfactory results. The advantage is that the simplified approach allows for studying a set of alternative scenarios within a relatively short period of time. At the same time, the authors acknowledge that for hourly modeling, it seems more appropriate to use an optimization approach.

The authors [29] consider it proven that existing models are insufficient for examining grids with a high percentage of VRE, and there arises a problem of adequately accounting for short-term temporal changes due to the complexity of the model and computational costs.

Given the rapid growth of renewable energy production and considering the need for widespread use of energy storage to ensure grid flexibility and stability, the authors [30] dedicated their review to two types of energy storage systems. They differentiate between storage systems that use direct and reverse conversion for energy storage (GIES) (for example, heat and electricity) and those that do not. From their perspective, long-term models of power systems are limited to considering energy systems with one type of storage without accounting for GIES. The authors believe that:

- Existing models are inadequate for grids with a high percentage of Variable Renewable Energy (VRE) when using storage systems without considering GIES;

- There is a problem with integrating short-term time changes into long-term models of power systems due to the complexity of the model and computational costs.

The key feature of the new structure proposed by the authors is the inclusion of "agent-based modeling" of consumer behavior, which allows reducing the number of scenarios for the operation of Variable Renewable Energy (VRE).

1.4 Transitioning Ukraine's Energy Sector to Variable Renewable Energy (VRE) by 2050

The works [31–40] are dedicated to the issue of Ukraine's energy sector transition to renewable energy sources by 2050.

"The mission of Ukraine's Energy Strategy until 2050 [31] is to create conditions for sustainable development of the national economy by ensuring access to reliable, resilient, and modern energy sources. By 2050, the energy sector should be as close as possible to climate neutrality. This will entail having clean energy, overcoming energy poverty, developing an innovative and decentralized energy system, ensuring the full functioning of national energy markets, and integrating them into international ones. The key principles of Ukraine's Energy Strategy are economic viability, environmental friendliness, accessibility, social justice, and market orientation. The Energy Strategy will be based on targeted economic development indicators in line with the National Economic Strategy for the period until 2030. It will also be aligned with Ukraine's international commitments, primarily within the framework of the Association Agreement between Ukraine and the EU and the Paris Climate Agreement." Therefore, according to the authors [31], the goals of Ukraine's Energy Strategy until 2050 are:

- Achieving maximum level of climate neutrality.
- Maximizing the reduction of coal usage in the energy sector.
- Updating and modernizing the energy infrastructure.
- Increasing resource efficiency in the energy sector.
- Comprehensive integration with the European Union markets and efficient functioning of domestic markets.
- Ensuring the energy sector's self-sufficiency considering economic viability.
- Developing alternative energy sources, new products, and innovative solutions in the energy sector.

The energy sector of Ukraine is in the process of transitioning from energy production based on fossil fuels to sustainable energy production from renewable sources [34]. This transition is accompanied by significant changes in national legislation, technical infrastructure, and integration into global processes related to the development of renewable energy sources and climate change mitigation. These processes include the Association Agreement with the EU, accession to the Energy Community and the Paris Agreement, as well as the implementation of the EU's Third Energy

Package and others. Three scenarios for the development of the energy sector by 2050 with increased targets for renewable energy sources (RES) have been proposed.

1. Base scenario.
2. Climate action scenario.
3. High ambition scenario for 100% RES.

The base scenario [34] is a continuation of the existing practice in the development of the energy sector. According to this scenario, there is a dominant role of energy production from fossil fuels, but some progress is also expected in the field of renewable energy sources and energy efficiency development in line with Ukraine's current commitments. At the same time, GHG emissions should not exceed 40% in 2050 compared to the current level, without surpassing the current level of intended nationally determined contributions (INDCs) submitted in accordance with the Paris Agreement.

The Climate Action scenario [34] is associated with Ukraine's climate commitments under the Paris Agreement and the Association Agreement with the European Union, specifically aiming for a 70% reduction in emissions by 2050 compared to the 2015 level. It is envisaged that these climate commitments can be achieved through a proportional increase in renewable energy sources, energy consumption reduction, and a restructuring of the energy balance of fossil fuels in the proportion of 13%/45%/12%. This scenario includes a reduction in the use of fossil fuels, especially coal, for energy production.

The High Commitment to 100% RE scenario [34] is linked to the "maximum feasible efforts" variant towards developing renewable energy sources and reducing energy consumption. It sets indicative targets for a 100% share of renewable energy sources in the sectors of electricity generation, heat, and transport. At such a level of RE utilization, there are some critical constraints regarding the stability of the energy system, the necessary investment volume, the overall potential of renewable energy sources available in Ukraine, and the general economic conditions during the modeling period.

Energy Strategy [33] includes several scenarios for security and economic development, depending on the end of the war, GDP growth rates, demographic situation, and Ukraine's integration into the European Union. Ukraine has the potential by 2050 to increase wind generation capacity to 140 GW, solar to 94 GW, energy storage to 38 GW, nuclear generation to 30 GW, combined heat and power (CHP) and bioenergy capacities to 18 GW, hydro generation to 9 GW. Overall, the investment opportunities for new energy capacities amount to $383 billion. Specifically, wind generation is $134 billion, solar is $62 billion, hydrogen technologies are $72 billion, energy storage is $25 billion, nuclear generation is $80 billion, and transmission systems are $5 billion, while hydroelectricity is $4.5 billion.

During the Ukraine Renewal Conference held in London, Energy Minister Herman Halushchenko presented Ukraine's Energy Strategy until 2050 [35]. The main goal of the strategy is to create a European energy hub in Ukraine. Our underground storage facilities can store up to 30 billion cubic meters of gas, which will guarantee the energy security of the European Union.

Even the brief overview of works dedicated to improving forecasting methods, modeling operating modes, and updating energy systems highlights the vast array of issues in this area and confirms the conclusion about the relevance of updating existing and developing new software-information systems for modeling modern complex energy systems. Therefore, the models proposed in this paper are aimed at:

- Addressing specific problems that arise during the assessment of the current state and quality of scenarios for forecasting the development of energy systems;
- Solving, generalizing, or simplifying some of the tasks listed above.

1.5 Modeling the Diffusion of Innovative Technologies

Energy systems are a particular case of large economic systems, whose development has been and continues to be the subject of numerous scientific studies. The presented analysis of works in this direction [41–44] allows us to conclude that there are patterns in their development. On the other hand, it is necessary to account for uncertainties caused by random unpredictable influences on the usual course of events [45]. In this context, we can only talk about temporary and conditional trends, which, while being trends over a relatively short period of time, can give way to opposite trends and thus may be part of a long-term cycle.

The forecasting of the development of complex economic systems is based on three main theoretical directions:

- The theory of economic equilibrium [46], which focuses on the study of stable state parameters, the causes of their disruption, and the mechanisms for their restoration.
- The theory of economic growth, which aims to determine the conditions for sustainable, balanced growth and development—Walter Rostow's stages of economic growth theory [47].
- The theory of economic cycles or business cycle theory [48], which explains the reasons for fluctuations in economic activity over time.

All three theories do not contradict but complement each other.

The theory of cycles is a systemic theory that studies the patterns in the formation of cycle structures in various types of processes. This scientific direction synthesizes scientific knowledge from the perspective of studying the temporal patterns of large systems. Systems develop cyclically, undergoing oscillatory processes. With a period of several years or decades, a growth phase is followed by a decline phase. Despite the oscillations, the overall development trend is characterized by growth. The theory of cycles, also known as Kondratiev wave theory, forms the basis of innovation theory and numerous innovative models of the diffusion of new technologies and their life cycle parameters.

Fundamental Model of Innovation Diffusion

The return from technology (such as the total number of users, installed capacity of power-generating technology, etc.) or the capital invested in it [49] occurs unevenly.

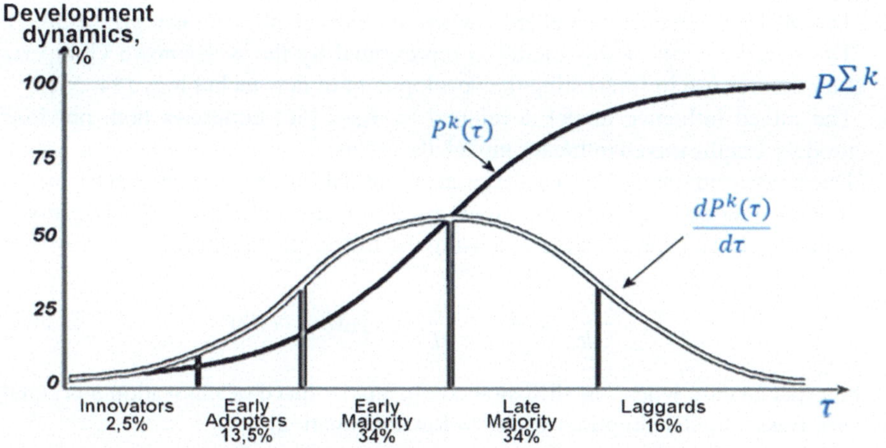

Fig. 1.1 Life cycle of an innovative technology

The general rule is that the return gradually increases, reaching a peak somewhere around the middle of its lifecycle, after which it decreases, resembling the graph of a normal distribution (Fig. 1.1).

The lower return at the beginning of the period is associated with the natural limitations inherent in any new resource and its initial development, and at the end, with the process of physical aging and competition from new products and technologies.

The founder of the innovation cycle theory, Kondratiev, studied the overall dynamics of cumulative values—capital and population—and their relationship with technological development. As he asserts [48], he was able to prove that the law governing the change of cumulative elements is expressed by a differential equation, which was later reformulated in terms of the introduction of innovative technologies and is considered the fundamental model of innovation diffusion [50]:

$$\frac{dN(t)}{dt} = g(t)(M - N(t)),$$

where $N(t)$—cumulative number of users at time t, who have adopted and are using the technology; M—maximum potential number of users; $g(t)$—diffusion coefficient (speed) that reflects the probability that potential users will adopt innovations within a small interval around time t. The value of $g(t)$ depends on characteristics of the diffusion process such as the type of innovations, communication channels, time, and characteristics of the social system.

Depending on the formula for the diffusion coefficient $g(t)$, three models of innovation diffusion were proposed [51]:

1. The external influence model [52], where the diffusion coefficient $g(t)$ is a coefficient or speed of innovation diffusion p;

2. The internal influence model [53], where the diffusion coefficient $g(t) = qN(t)$. The specific form of this model is represented by the well-known Gompertz function for use in forecasting the development of new technology [54, 55];
3. The mixed influence model developed by Bass [56] combines both previous models. For the mixed influence model, the diffusion coefficient $g(t) = p + qN(t)$. Due to its comprehensive consideration of internal and external influences, mixed influence models are most commonly used in research [51, 57, 58]. The mixed influence model can be expressed using the following equation:

$$\frac{dN(t)}{dt} = \left(p + \frac{q}{M}N(t)\right)(M - N(t)) \tag{1.1}$$

In a special case where the diffusion coefficient or speed of innovation p is equal to zero, Bass's model simplifies to the following equation:

$$\frac{dN(t)}{dt} = \frac{q}{M}N(t)(M - N(t)).$$

This model is called the logistic model.

Both models—the Bass model and the logistic model—yield S-shaped curves of cumulative user count. By definition, the S-shaped diffusion (Fig. 1.1) initially increases at an accelerating rate as the cumulative number of users grows over time. Eventually, the curve reaches an inflection point, and the diffusion rate begins to decrease. Finally, diffusion reaches the saturation level M.

E. Rogers [59] believes that the majority of innovation adoption curves (Fig. 1.1) by society members resemble a normal distribution. The curves demonstrate the speed and stages of innovation diffusion within the community. Rogers gave his own name to each segment, based on the mean and standard deviation:

- Innovators—2.5%;
- Early Adopters—13.5%;
- Early Majority—34%;
- Late Majority—34%;
- Laggards—16%.

All the discussed models assume a constant maximum potential number of users, M. However, in reality, M is not a constant. Therefore, models are needed in which the number of potential users is a function of time $M(t)$.

In research dedicated to innovations and innovative development, the concept of technological paradigm gained particular popularity, stemming from the technological-economic paradigm proposed by a group of Western economists in the 1970s–1980s [60].

The life cycle of each technological paradigm [61] lasts approximately 100 years on average. It encompasses the life cycles of technologies, each with its own duration. Detailed studies have shown that the life cycle of a technology undergoes two peaks, two upward waves, during its existence. The first peak occurs at the beginning of the technological paradigm's development and is driven by technological, internal factors, triggered by the regularity of new technology offerings when the paradigm makes its way into a foreign socio-economic environment. The second peak occurs at the beginning of the second half of the life cycle, when economic relations in society have transformed sufficiently to accept the technological innovations proposed by this paradigm. This peak is driven not by technological but by economic factors, external to the development of the technological production base, and reflects society's readiness to adopt corresponding innovations and the systematic growth of societal demand for them. These two peaks in the development of technological paradigms—endogenous and exogenous—overall reflect the quantitative dynamics of various step-cycle processes themselves. The model of the quantitative dynamics of technological paradigms, known as the Grubler-Fetisov hypothesis, allows for fairly accurate forecasting of transitional and crisis periods in the development of techno-economic macrosystems and individual technologies. Overall, the replacement of technological paradigms occurs according to a law that is mathematically described by a generalized logistic curve.

Generalizing Bass's model (1) into a generalized logistic curve and taking into account the shortcomings listed in [50] allowed for the formulation of dynamic models of cyclical development of power generation systems, a method for economic-technological forecasting of optimal functioning and development parameters of the energy system, and a mathematical model for long-term technological renewal of the consumption structure and generating capacity of the power system of Ukraine in Chap. 2 of this book.

1.6 Modeling the Energy System as a Complex Hierarchical Quasi-Dynamic System

The analysis presented in [62] includes a list of typical tasks for the development of the energy system, which includes:

1. Forecasting consumption and its patterns.
2. Investigating long-term trends in energy technology development.
3. Selecting a rational structure.
4. Analyzing efficiency, both of individual components and the entire energy system.
5. Selecting a set of rational development options for the energy system.

In addressing these tasks [62], "models of a unified energy system are created, a rational structure of generating capacities of the unified energy system and energy associations by types of equipment is selected, a rational structure of generating capacities of the unified energy system and individual energy associations, as well as the prospective equipment needs, are substantiated." At the next level, a set of optimal development and upgrading options for the components of energy associations for the future is selected. Various mathematical models are used for the analysis of energy systems, both existing and newly created. As of 2019, there were 146 known energy modeling systems worldwide [62, Table 2.1, p. 83]. The main indicators of these modeling systems are presented in Table A.1. They are classified [62] "into four target groups: M—Modeling, I—Investing, U—Managing, and U + I—Managing + Investing." Table A.2 contains a list of economic equilibrium modeling systems for energy complexes [62], Table A.3—lists mathematical problem solvers [62], and Table A.4—lists energy equilibrium modeling systems [62] as of 2019.

The majority of studies have concluded that developing a single tool that equally meets the variety of requirements for energy system models seems unpromising. Therefore, this work proposes a method and economic-mathematical models aimed at addressing specific problems that arise in assessing the current state and quality of scenarios for the development of energy systems.

Table 1.1 presents the known models, the modification and generalization of which allowed for the development of a new method and improved models.

Based on the modification of the fundamental model of innovation diffusion and the generalization of the mixed influence model of new technology diffusion, a new dynamic model of cyclical development of generation systems was developed. This model, in turn, allowed for the development of a new method of economic-technological forecasting of optimal operation and development parameters of the energy system, as well as a new model of long-term technological renewal of consumption structure and generating capacity of the power system of Ukraine. Generalizing the classical model of dynamic system management led to the development of a new model of hierarchically controlled quasi-dynamic energy system. Specific cases of this model include the developed model for minimizing the weighted average cost of electricity and the model of a multi-nodal integrated energy system. The software modules for numerical implementation of these mathematical models were integrated into the developed problem-oriented software-information complex.

Table 1.1 Known and new/improved models

Known models	New and improved models	
The fundamental model of innovation diffusion	Method of economic-technological forecasting for optimal operation and development parameters of the energy system	
	Dynamic models of cyclical development of generation systems	
Mixed influence model of diffusion of new technologies	Long-term technological renewal model of consumption structure and generating capacity of ukraine's integrated power systems	
The classic formulation of dynamic system management problem	Model of hierarchically managed quasi-dynamic energy system	
	Model for minimizing the weighted average cost of electricity	Model of a multi-nodal integrated energy system

The following is a comparison of the developed new software-information complex [63, 64], proposed in this work, with the most well-known software-information tools for modeling energy systems in Ukraine:

- PLEXOS [65]—developed by Energy Exemplar Pty. Ltd. (energyexemplar.com/plexos)—a commercial system used by ENTSO-E, as well as some participants in the electricity market, including traders and DSOs, as well as NPC "Ukrenergo".
- ANTARES SIMULATOR [66]—developed by RTE France (antares-simulator.org)—freely distributed, used by ENTSO-E and some ENTSO-E DSOs, as well as NPC "Ukrenergo".
- FlexTool [67]—developed by VTT Technical Research Centre of Finland Ltd (irena.org/energytransition/Energy-System-Models-and-Data/IRENA-FlexTool)—freely distributed, used by IRENA.

Each of these modeling tools has its own characteristics, so the solutions usually do not coincide. A comparison of the mentioned models with the software-information complex proposed in this study [63, 64] and described in Chap. 3 is presented in Table 1.2.

As can be seen from Table 1.2, the software-information complex proposed in this study has several advantages:

- Adequate modeling of the operating modes of hydro units of hydroelectric power plants (HPPs) and power units of thermal power plants (TPPs) and nuclear power plants (NPPs) in Ukraine.
- The ability to compare multiple models and optimization results at different stages of the life cycle with the same input data.

The methodology of systems research in the energy sector today [62] includes the following fundamental principles:

- Study of the nature of the investigated energy systems.
- Investigation of the fundamental properties of energy systems, which evolve with the development of these systems.
- Development and updating of models and research methods for energy systems to adequately account for the changing properties and development trends.
- Consideration of not only the specificity of models but also a large number of criteria and constraints, the conflicting nature of interests of various stakeholders.
- Identification of controlling influences to ensure necessary reliability, resilience, and manageability.

The general scheme of energy research (Fig. 1.2) includes:

- Application of mathematical models.
- Consideration of uncertainty in information about future energy development conditions: rates of growth in fuel and energy resource demand, prospects of scientific and technological progress, and environmental constraints.
- Consideration of territorial distribution of energy production capacities and levels of electricity consumption in territorial nodes.

Table 1.2 Key differences between well-known software-information tools for energy system modeling and the software-information complex proposed in this study

Parameter	IZE NAS of Ukraine	Plexos	Antares simulator	Flex tool
Limits of export/import per day, MWh	+	+	+	+
Limits of export/import per hour, MWh	+	+	+ / −	+
The possibility of maintaining a zero balance of export/import per day	+	+	+	+
Taking into account the price of Export/Import of electricity for each day, USD/MWh	+	+	+	+
Adequate modeling of hydro units of the pumped storage power plants (PSPP) in Ukraine	+	−	−	−
Adequate modeling of TPP and NPP units in Ukraine	+	+ / −	+ / −	−
Wind power plant (WPP) and solar power plant (SPP) modeling according to annual power generation profiles	+	+	+	+
Battery storage systems modeling (energy stored ≥ generation)	+	+	+	+
HPPs modeling (minimum water release, constraints on power variation, daily production limits)	+	+	+	−
Multi-node energy system modeling	+	+	+	+
Capability to compare multiple models	+	−	−	+ /−
Ability to compare optimization results at different stages of the life cycle	+	−	−	+ /−
User interface	+	+	+	+

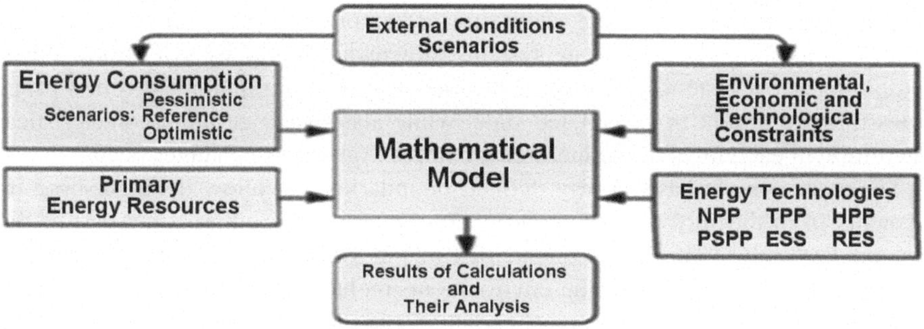

Fig. 1.2 General scheme of energy research

- Investigation and consideration of constraints on updating structural elements of the fuel and energy complex.
- Processing and analysis of results obtained from the application of mathematical models.

Research is conducted taking into account possible energy consumption scenarios: pessimistic, reference, optimistic. The technological and economic efficiency of each technology is considered and evaluated. The availability of primary energy resources is taken into account both at present and in the future.

Optimization based on economic criteria allows for the identification of long-term energy development trends that are most likely to be implemented in the future. Theoretical analysis indicates that solving the global optimization problem for minimizing costs aligns with solving local problems for maximizing profit, provided that relevant constraints are considered in these tasks. Given this, when constructing a system of mathematical models for studying the prospects of energy development, the energy system as a whole or its individual components, modeled at their respective hierarchical levels, are considered as a controlled quasi-dynamic system. The trajectory of this system, under the influence of permissible optimal controls, represents a sequence of states of optimal energy and economic equilibrium [68]. In modeling, each element of the system is treated as a complex entity, considering its specifics. The system consists of numerous elements that form its structure and determine its behavior under certain conditions. An element is a component of the system that is indivisible in terms of performing a specific function. Each system is an element of a higher-order system, and any element can be considered a relatively autonomous system composed of lower-order elements. Thus, within a system, an element is the minimal unit capable of relatively independent execution of a particular function.

Most of the tasks in system analysis involve finding the optimal value of an objective function while considering certain constraints. In practice, multiple objectives are often formulated, and a hierarchy is established among them, formalized through the construction of multi-objective models. Due to the hierarchical nature, each system or subsystem must occupy a specific level in the hierarchy, meaning it is an element of a higher-order system (upward hierarchy extension) while simultaneously consisting of subordinate systems (downward hierarchy extension). When combining into a system, each level of the hierarchy acquires qualitatively new properties that are absent in an isolated state, while also losing certain characteristics. Therefore, at each level of the hierarchy, complex qualitative changes occur.

The system must have clearly defined boundaries that allow distinguishing its elements from the surrounding environment, i.e., everything that is not part of the system. Precisely defining the boundaries of the system also enables the investigation of its interaction with the environment, including the order of information exchange, technologies, resources, the system's impact on the environment, and the environment's influence on the internal structure of the system.

The ability of a system to develop is a necessary condition for its viability, which is ensured through self-improvement and self-preservation. System development is the transition from one state to another while preserving its essential systemic characteristics. The state of the system is described by a set of parameters that most fully characterize it at any given moment. The process of system development involves maintaining dynamic equilibrium. On one hand, dynamism means constant development, adaptation to changes in the environment, improvement, and transition to higher hierarchical levels. On the other hand, during dynamic development, the system must maintain equilibrium as the ability to preserve its fundamental systemic characteristics in the absence of external disturbances or under constant such actions.

The classic formulation of the dynamic system control problem (DSC) [69, 70] assumes that control decisions are made by a single entity based on a single optimality criterion, which can be expressed as:

$$\mu = \sum_{\tau=1}^{T} g(\Omega(\tau), u(\tau), \xi(\tau)) \rightarrow \max,$$

where at the moment of time τ:

- g—the optimality criterion value,
- Ω—state vector of the dynamic system,
- u—control action vector τ,
- ξ—vector of random external actions.

The works on optimal control theory [71, 72] are dedicated to solving this problem in various ways.

Based on the considerations outlined and considering that the classical model of dynamic system control is insufficient for modeling a complex system like the energy system, it is generalized into *an economic-mathematical model of hierarchically controlled quasi-dynamic energy system* with $r \in R$ levels of administrative-territorial hierarchy and sectoral (subsectoral) infrastructure that is detailed in terms of its technological $k \in K$ structure. The problem of hierarchical control of such a quasi-dynamic system is formulated in Chap. 2 of this book.

1.7 Modeling a Multi-node Integrated Energy System with Integer Variables

A special case of the model of a hierarchically controlled quasi-dynamic system is the developed economic and mathematical model of a multi-node integrated energy system [70] with integer variables that optimize the structure and operating modes of the generating, reserve and storage capacities of energy systems operating in a single synchronous zone of continental Europe ENTSO-E.

Based on the requirements outlined in [73–80], the technological conditions for the parallel operation of the power systems of Ukraine with neighboring countries within ENTSO-E [73] have been considered. The basic scenarios for the development of generating capacities were analyzed, and influential factors regarding the readiness of Ukraine's power system for integration with the Continental European power system ENTSO-E were determined. These form the basis for modeling the synchronous operation of Ukraine's power system with the power systems of neighboring countries. Specific features of the operation of power systems in neighboring countries of Ukraine, members of the European Union—Hungary, Slovakia, Poland, Romania—and Ukraine's power system were discussed, allowing for the identification of prerequisites for synchronization considering the limits of cross-border electricity exchange, geographical location, the presence and capacity of interstate transmission lines, and the magnitude of maximum permissible import and export volumes. Overall, this allowed for substantiating the inevitability and economic feasibility of the parallel operation of Ukraine's power system with the Continental European power system ENTSO-E, which, under the introduction of new dispatching regimes and maneuverable storage and reserve power generation capacities, will ensure stability, energy security, and dynamic and static stability of Ukraine's power system operation.

Model [70] belongs to the class of optimal dispatch problems for power generation capacities in power systems with the criterion of minimizing production costs and import/export of electrical energy. In the most general form, the model can be described as follows:

$$\sum_{r}^{R}\sum_{\tau}^{T}\left\{\sum_{g}^{G}\left(P_{r\tau g}^{Gen}c_{r\tau g}^{Gen}+S_{r\tau g}^{Start}c_{r\tau g}^{Start}\right)+\sum_{k}^{R}P_{rk\tau}^{Imp}c_{k\tau}^{Imp}(1-a_{kr}^{Transf})\right\}\rightarrow \min,$$

Production—Consumption—Import—Export Balance:

$$\sum_{g}^{G}P_{r\tau g}^{Gen}+\sum_{k}^{R}\left(P_{rk\tau}^{Imp}-P_{rk\tau}^{Exp}\right)=D_{r\tau};\ \forall\tau\in T;\ \forall r\in R.$$

The power flow balance includes:

- constraints on the transmission line capacity (Net Transfer Capacity):

$$P^{Exp}_{rk\tau} \le F^{NTC}_{kr}; P^{Imp}_{rk\tau} \le F^{NTC}_{kr}; \forall \tau \in T; \forall r \in R; \forall k \in R(k = r).$$

- possible constraint on import balance—the import from node r to node k must equal the export from node k to node r:

$$P^{Imp}_{rk\tau} - P^{Exp}_{rk\tau} = 0; \forall \tau \in T; \forall r \in R; \forall k \in R(k = r).$$

Here's a breakdown of the symbols used in the model:

- T—period (time in hours);
- $R(K)$—set of nodes;
- G—set of generating capacities;
- P—power;
- c—cost;
- α—loss coefficient;
- D—consumption;
- F—maximum transmission capacity of interconnecting power lines.

For more detailed information, refer to mathematical model [70, 81] described in Chap. 2.

1.8 Model for Minimizing the Weighted Average Cost of Electricity Generation and Storage

One of the key factors [82–85] in assessing the economic efficiency of modern electricity generation technologies is the average cost of production over the study period. Apart from the primary investment and operational components, the cost depends on the utilization of the installed capacity of the power plant. This factor is largely determined by the consumption schedule, which, in turn, affects the load schedule. Smoothing the load schedule requires significant technological and economic resources and the use of generation technologies in a load-following mode.

The model [86], which is described in detail in the second chapter, allows for the assessment of the potential reduction in the average hourly cost of electricity generation and, based on forecast daily load profiles, addresses the task of predicting the optimal operating modes for the system's generation and storage capacities. Utilizing the obtained solution ensures the minimization of the hourly cost of electricity produced by Ukraine's Unified Energy System (UES).

1.9 Long-Term Technological Renewal of Energy Systems Components Modeling

The study of the development and improvement of energy systems is a relevant scientific problem that receives significant attention [87, 88]. In the work [89], an enhancement of the mathematical programming model is proposed, based on the modification and development of a family of hierarchical stepwise optimization models for the sustainable development trajectories of energy systems [90], and a stochastic quasi-dynamic model of the economic and technological impact of the life cycle of innovative technologies [91]. The main distinction of the proposed model [89] is the explicit consideration of the economic and technological indicators of national economic development and production, represented as quasi-dynamic functions with discrete stochastic variables. The calculations used actual and forecasted values of these functions at the respective steps of the modeling horizon. The modification and development of these models aim to adapt them for researching the directions and optimal parameters for the technological renewal of energy system components operating within energy unions, in the long-term perspective and considering their future prospects.

The assessment of relative costs for modernizing energy systems that use different technologies is a complex task [92–94]. The results of this assessment are highly location-dependent. For example, coal, given the current situation, is and will likely remain economically attractive in countries where carbon emissions are either taxed at low rates or not taxed at all. Gas can also compete with base load generation, especially when using combined cycle plants. NPPs [82, 85] are expensive to build but relatively cheap to operate. The costs of waste disposal and decommissioning are usually fully included in the operational costs. In many locations, nuclear energy successfully competes with fossil fuels as a means of electricity production. Considering the social, health, and environmental costs of using fossil fuels increases the competitiveness of nuclear energy. Life cycle assessments of technology show that nuclear energy is an economically viable source of electricity production, combining the advantages of safety, reliability, and very low greenhouse gas emissions. Existing NPPs operate reliably with a high degree of predictability. The operational costs of NPPs are lower than those of all competitors using fossil fuels and have a very low risk of operational cost inflation. Experience shows that the lifecycle of a NPP is 60 years and can be extended further in the future. The main economic risks for existing NPPs lie in competition with subsidized renewable energy, whose production volumes are difficult to predict, and with the production of cheap gas, especially considering the political risks of increasing specific taxation on nuclear energy.

The method of economic and technological forecasting of optimal functioning and development parameters of the energy system, along with the model for long-term technological renewal of the consumption structure and generating capacity of Ukraine's Integrated Power System (IPS), which is a component of this method, is designed to study short- and long-term scenarios. The model is a specific case of the model of a hierarchically controlled quasi-dynamic system [90] and the quasi-dynamic model of the economic and technological impact of the life cycle of innovative technologies [91]. The primary distinction of the proposed model is the explicit consideration of the economic and technological indicators of national economic development and production, represented as quasi-dynamic functions with discrete stochastic variables. The model is presented as a hierarchy of scenarios. At the top level of the hierarchy, aggregated energy supply technologies are employed. The system state matrix reflects the structure of generation, supply, and consumption volumes at each step of the modeling horizon. All utilized technologies participate in ensuring the balance of supplied and consumed energy at each step. The main constraints of the model include maintaining the balance of total capacity, maneuverable capacity, and the volumes of generated, supplied, and consumed energy, provided that all parameters belong to the set of possible states. A measure of discrepancy between the vectors of supplied and consumed energy is introduced. The input information for modeling [95–106] includes:

- Target sequence of annual energy consumption;
- Initial state of the total supply vector;
- Vector of integral differences of LACE–LCOE parameters for each of the aggregated technologies involved in the calculation;
- Functional of economic-technological impact, which includes:

  a. Regional purchasing power coefficient,
  b. Technology efficiency,
  c. Final cost coefficient,
  d. Economic growth rate.

The forecasted cost of technology components is also taken into account: investments, fixed operating costs, variable operating costs. The formulation and description of the model are provided in Chap. 2.

1.10 Formulation of the Goal and the Statement of Research Tasks

The analysis of scientific and other informational sources has revealed that the development of theoretical principles, methods, and tools for assessing the efficiency of functioning and development of energy systems, as well as mathematical methods and models for determining the possibilities and limitations of updating elements of both autonomous and integrated energy systems, has been the focus of work by many Ukrainian and foreign scientists.

The relevance of this research is driven by key trends in the energy sector. Optimizing the structure of energy systems with renewable energy sources (RES) is crucial for achieving sustainable development goals and mitigating climate change. Progress in technologies and energy storage provides new opportunities for optimal integration of RES into existing and new energy systems. Models and tools that take into account economic and technological aspects of RES integration are vital for the reliability and resilience of energy systems. They are crucial for maintaining network stability and making strategic decisions in the energy sector. Optimizing energy systems using RES contributes to the formation of effective scenarios for a sustainable energy future. Efficient utilization of energy storage technologies is a key aspect of optimizing energy systems with a high share of RES.

Given the importance of optimizing energy systems with renewable energy sources (RES), the task of developing new and improving existing methods, economic-mathematical models, and software tools to investigate directions and optimal parameters for technological development in energy systems is highly relevant. This research is dedicated to addressing this scientific challenge.

The above allows us to formulate the purpose and objectives of the research.

The purpose of the research is to develop new and improve existing methods, economic-mathematical models, and software tools for optimizing the composition and operating modes of generating, reserve, and storage capacities of energy systems in the context of rapid growth of renewable energy capacities worldwide and in Ukraine.

The following tasks were formulated and solved in the work:

- Explored the possibilities of optimized dispatching of energy system components, the limits of acceptable values of parameters of generating, reserve, and accumulating capacities ensuring stable balanced operation of the integrated energy system.
- Improved economic-mathematical models for determining the technical–economic efficiency of operation of individual electricity generation objects throughout their life cycle.
- Developed a method of economic-technological forecasting of parameters for optimal operation and development of the energy system.
- Developed new and improved existing economic-mathematical programming models for researching directions and determining optimal parameters for technological development of energy system elements.
- Conducted computational experiments using the developed method, models, and software tools to identify prospects for short- and long-term development of Ukraine's power systems.

1.11 Conclusions

The possibility of sustainable economic development of the national economy is largely determined by the reliability and efficiency of energy supply systems. At the same time, with the increasing variability of the global economic system, there is a rise in environmental requirements for technological processes, including those related to the extraction, conversion, and consumption of energy resources and carriers. Currently, there are no universal economically viable methods for harmoniously addressing the issue of meeting humanity's energy needs while ensuring no negative impact on the environment.

The development of theoretical principles, methods, and tools for assessing the efficiency of operation and development of energy systems, as well as mathematical methods and models for determining the capabilities and limitations of updating elements in both autonomous and integrated energy systems, is the focus of work by many Ukrainian and foreign scientists.

Existing studies either consider models of integrated energy systems with a single integrated optimization criterion or optimize the functioning of individual nodes or elements of the energy system. These approaches do not allow for simultaneously achieving detailed optimization of both the entire energy system and the optimized use of its individual technological components.

Therefore, an important task is the improvement and development of new mathematical programming models for investigating the directions and optimal parameters of technological updates for elements of energy systems operating within energy unions. Incorporating economic and technological indicators of national economic development and production into these models, represented as quasi-dynamic functions with discrete stochastic variables, allows for studying the optimal volumes of technological updates in integrated energy systems.

The existing situation highlights the relevance of developing appropriate new tools and improving existing methods that combine the advantages of the mentioned approaches. This scientific task is the focus of the presented work. The software-information complex proposed in this study has the following additional advantages:

- adequate modeling of the operation modes of hydro units in HPPs and power units in TPPs and NPPs in Ukraine;
- the ability to compare multiple models using the same input data;
- the ability to compare optimization results at different stages of the lifecycle.

References

1. Kulyk, M., Horbulin, V., Kyrylenko, O.: Conceptual approaches to the development of energy of Ukraine (analytical materials): монографія. Kyiv : Instytut zahalnoi enerhetyky NAN Ukrainy, 78 (2017)
2. Stogniy, B.S., Kyrylenko, O.V., Prakhovnyk, A.V., Denysiuk, S.P.: Evolution of intelligent electrical networks and their prospects in Ukraine. Tech. Electrodyn. **5**, 52–67 (2012)
3. The international renewable energy agency (IRENA). Systemic Innovation. https://www.irena.org/Energy-Transition/Innovation/Systemic-Innovation. Date of Access: 09 Mar 2023
4. Oliynyk, D.: Smart energy: global trends and domestic realities. Viche **21**, 46–48 (2015)
5. Shlyaha, O.V.: Trends in the development of innovative processes in Ukraine and their impact on the economy under modern conditions. Econ. Bull. Donbass **3**(17), 139–144 (2009)
6. Renewable Technology Innovation Indicators: mapping progress in costs, patents and standards. IRENA—International Renewable Energy Agency. https://www.irena.org/publications/2022/Mar/Renewable-Technology-Innovation-Indicators. Date of Access: 27 July 2023

7. Obleshchenko, A.D., Postol, Y.O.: Key trends in the development of energy management. In: II All-Ukrainian Scientific and Practical Internet Conference April 5–25, 2021: Proceedings of the II All-Ukrainian Scientific and Practical Internet Conference, April 5–25, 2021, Melitopol. Melitopol, pp. 8–9 (2021)
8. Pivnyak, H., et al.: Traditional and non-traditional energy supply systems for urbanized and industrial areas of Ukraine: Monograph. National Mining University, 333 pp (2013)
9. Veremiychuk, Y., et al.: Optimization of integrated energy supply systems for consumers. In: Institute of Technical Thermophysics, NAS of Ukraine, Kyiv, 186 pp (2020)
10. Kulyk, M., et al.: Energy Efficiency and Forecasting Energy Consumption at Various Hierarchical Levels of the Economy: Methodology, Forecast Evaluations up to 2040. Naukova Dumka, Kyiv, 235 pp (2021)
11. Maliarenko, O.Y., Teslenko, O.I., Maistrenko, N.Y. Mathematical model for forecasting the electricity consumption for the country and the industrial sector by months of the year. In: Achievements of ukraine and eu Countries in Technological Innovations and Invention, pp. 263–294 (2022). https://doi.org/10.30525/978-9934-26-254-8-9. Date of Access: 27 July 2023
12. Chen, B., et al.: Pathways for sustainable energy transition. J. Clean. Prod. **228**, 1564–1571 (2019). https://doi.org/10.1016/j.jclepro.2019.04.372. Date of Access: 27 July 2023
13. Explore the energy system by technology or sector. Ukraine energy profile—Analysis—IEA. *IEA*. https://www.iea.org/reports/ukraine-energy-profile. Date of Access: 27 July 2023
14. Kudrya, S. (Ed.) Renewable Energy Sources. Institute of Renewable Energy, NAS of Ukraine, Kyiv, 392 pp (2020)
15. World Energy Outlook 2019—Analysis—IEA. *IEA*. https://www.iea.org/reports/world-energy-outlook-2019#introduction. Date of Access: 27 July 2023
16. Global Energy and Climate Model—Analysis—IEA. *IEA*. https://www.iea.org/reports/global-energy-and-climate-model. Date of Access: 27 July 2023
17. International Renewable Energy Agency (IRENA). Global Energy Transformation: A Roadmap to 2050. https://www.irena.org/publications/2019/Apr/Global-energy-transformation-A-roadmap-to-2050. Accessed: 27 July 2023
18. M. Geidl et al.: The energy hub: a powerful concept for future energy systems (2007). https://api.semanticscholar.org/CorpusID:14432048
19. Legal provisions of COM (2021) 350—Updating the 2020 New Industrial Strategy: Building a Stronger Single Market for Europe's Recovery—EU Monitor. EU Monitor—EU monitor. https://www.eumonitor.eu/9353000/1/j4nvhdfcs8bljza_j9vvik7m1c3gyxp/vlikpnsao4yk. Date of Access: 27 July 2023
20. Voropai, N., et al.: A multi-agent approach to electric power systems. IntechOpen—Open Science Open Minds | IntechOpen. http://www.intechopen.com/books/multi-agent-systems-modeling-interactions-simulations-andcase-studies/a-multi-agent-approach-to-electric-power-systems. Date of Access: 27 July 2023
21. Eutukhova, T., et al.: Methodological platform for determining energy efficiency indicators of a complex system. Syst. Res. Energy **2023**(2), 38–47 (2023). https://doi.org/10.15407/srenergy2023.02.038. Date of Access: 27 July 2023
22. Kaplin, M., Bilan, T.: Aggregated energy supply model according to product energy balance data structure. Syst. Res. Energy **2023**(2), 48–61 (2023). https://doi.org/10.15407/srenergy2023.02.048. Date of Access: 27 July 2023
23. Shulzhenko, S.: Generation unit commitment mixed integer linear model for simultaneous heat and electric daily load covering. Syst. Res. Energy. **2023**(1), 25–34 (2023). https://doi.org/10.15407/srenergy2023.01.025. Date of Access: 27 July 2023
24. Voropai, N., et al.: Simulation modeling of integrated multi-carrier energy systems. Simulation Modelling [working title]. 2021. https://doi.org/10.5772/intechopen.99323. Date of Access: 27 July 2023
25. Huang, B., et al.: A neurodynamic-based distributed energy management approach for integrated local energy systems. Int. J. Electr. Power & Energy Syst. **128**, 106737 (2021). https://doi.org/10.1016/j.ijepes.2020.106737. Date of Access: 27 July 2023

26. M. Fodstad et al.: Next frontiers in energy system modelling: a review on challenges and the state of the art. Renew. Sustain. Energy Rev. **160**, 112246 (2022). https://doi.org/10.1016/j.rser.2022.112246. Date of Access: 27 July 2023
27. Diangelakis, N.A., Pistikopoulos, E.N.: A multi-scale energy systems engineering approach to residential combined heat and power systems. Comput. & Chem. Eng. **102**, 128–138 (2017). https://doi.org/10.1016/j.compchemeng.2016.10.015. Date of Access: 27 July 2023
28. Pfenninger, S., Pickering, B.: Calliope: a multi-scale energy systems modelling framework. J. Open Source Softw. **3**(29), 825 (2018). https://doi.org/10.21105/joss.00825. Date of Access: 27 July 2023
29. Bottecchia, L., et al.: The potential of simulating energy systems: the multi energy systems simulator model. Energies **14**(18), 5724 (2021). https://doi.org/10.3390/en14185724. Date of Access: 27 July 2023
30. C.S. Lai et al.: A review on long-term electrical power system modeling with energy storage. J. Clean. Prod. **280**, 124298 (2021). https://doi.org/10.1016/j.jclepro.2020.124298. Date of Access: 27 July 2023
31. Ministry of Energy of Ukraine. Energy Strategy of Ukraine for the Period up to 2050. https://mev.gov.ua/reforma/enerhetychna-stratehiya. Date of Access: 27 July 2023
32. Dziak, O., Chepelyev, M., Podolyets, R., Tripolskaya, H., et al.: In: Ogarenko, Y., Aliyeva, O. (Ed.) Transition of Ukraine to Renewable Energy by 2050. ARTBOOK Publishing House, Kyiv, 88 p (2017). ISBN 978-617-7242-35-1
33. "Strategy for the Future: Ukraine as an Energy Hub to Help Europe Eliminate Dependence on Russia." Government Portal. Ministry of Energy of Ukraine. https://www.kmu.gov.ua/news/stratehiia-maibutnoho-ukraina-tse-enerhetychnyi-khab-iakyi-dopomozhe-ievropi-pozbutysia-zalezhnosti-vid-rosii. Accessed: 21 July 2023
34. Transition of Ukraine's Energy Sector to Renewable Energy by 2050. Renewable Energy Agency. https://rea.org.ua/wp-content/uploads/2019/12/aset_report_rea_ua_31082017.pdf. Accessed: 21 July 2023
35. Halushchenko, H.: Energy Strategy of Ukraine until 2050. Vector-VS. Available at: https://vector-vs.com/blog/view/energetichna-strategija-ukraini-do-2050-roku. Accessed 21 July 2023
36. Стратегія низьковуглецевого розвитку України до 2050 року. Разумков центр. https://razumkov.energy/files/docs/Proekt/LEDS_ua_last.pdf. дата звернення: 21 July 2023
37. Integration into the European Electricity Grid ENTSO-E. Ministry of Energy of Ukraine. Available at: https://www.mev.gov.ua/reforma/intehratsiya-u-yevropeysku-elektromerezhu-entso-e. Accessed 21 July 2023
38. New Law: On Energy Efficiency, Effective from November 13, 2021. Ministry of Energy of Ukraine. Available at: https://www.mev.gov.ua/storinka/enerhoefektyvnist-0. Accessed 21 July 2023.
39. Geyets, V.M., et al.: Energy strategy: projections (review). Nauka ta innovacii. **16**(1), 3–15 (2020). https://doi.org/10.15407/scin16.01.003. Date of Access: 21 July 2023
40. Litvak, O.A.: Green Energy as a Key Element of Low-Carbon Development: Global Trends and Prospects for Ukraine. In: Contemporary Trajectory of Scientific and Technical Progress in Ukraine and the World, pp. 80–112 (2021). Available at: https://doi.org/10.36059/978-966-397-247-3-3. Accessed 21 July 2023
41. Agajie, T.F., et al.: A comprehensive review on techno-economic analysis and optimal sizing of hybrid renewable energy sources with energy storage systems. Energies. **16**(2), 642 (2023). https://doi.org/10.3390/en16020642. Date of Access: 27 July 2023.
42. Ceglia, F., et al.: An energy, environmental, management and economic analysis of energy efficient system towards renewable energy community: the case study of multi-purpose energy community. J. Clean. Prod. **369**, 133269 (2022). https://doi.org/10.1016/j.jclepro.2022.133269. Date of Access: 27 July 2023
43. Seedahmed, M.M.A., et al.: A techno-economic analysis of a hybrid energy system for the electrification of a remote cluster in western Saudi Arabia. Alex. Eng. J. **61**(7), 5183–5202 (2022). https://doi.org/10.1016/j.aej.2021.10.041. Date of Access: 27 July 2023

44. Tawfik, H., et al.: Economic analysis of renewable energy systems. In: 2006 IEEE Long Island Systems, Applications and Technology Conference, Farmingdale, NY, USA, 5 May 2006 (2006). https://doi.org/10.1109/lisat.2006.4302663. Date of Access: 27 July 2023
45. Denisov, V.: Comparative assessment of control algorithms considering the randomness of external actions. In: Collection of Scientific Reports. Current Scientific Problems. Consideration, Decision, Practice, 7th ed. Wrocław, vol. 7, pp. 32–36 (2014)
46. Mas-Colell, A.: The theory of general economic equilibrium. Cambridge Core. https://doi.org/10.1017/CCOL0521265142. Date of Access: 27 July 2023
47. Jacobs, J.: Rostow's stages of growth development model (2023). In: ThoughtCo, thoughtco.com/rostows-stages-of-growth-development-model-1434564
48. Kondratieff, N.D.: The Static and the Dynamic View of Economics. OUP Academic. https://doi.org/10.2307/1883266. Date of Access: 27 July 2023
49. Short, W., Packey, D., Holt, T.: A manual for economic evaluation of energy efficiency and renewable energy technologies. National Renewable Energy Laboratory, U.S. Department of Energy Managed By Midwest Research Institute, p. 120 (1995)
50. Kijek, A., Kijek, T.: Modelling of innovation diffusion. Oper. Res. Decis. **3–4**(20), 53–68 (2010)
51. Hanssens, D.M., Mahajan, V., Peterson, R.A.: Innovation diffusion: models and applications. J. Mark. Res. **22**(4), 468 (1985). https://doi.org/10.2307/3151593. Date of Access: 27 July 2023
52. Fourt, L.A., Woodlock, J.W.: Early prediction of market success for new grocery products. J. Mark. **25**(2), 31–38 (1985). https://doi.org/10.1177/002224296002500206. Date of Access: 27 July 2023
53. Mansfield, E.: Technical change and the rate of imitation. Econometrica **29**(4), 741–766 (1961). https://doi.org/10.2307/1911817. Date of Access: 27 July 2023
54. Wildt, A.R., Linstone, H.A., Sahal, D.: Technological substitution: forecasting techniques and applications. J. Mark. Res. **14**(3), 425 (1977). https://doi.org/10.2307/3150798. Date of Access: 27 July 2023
55. Nutt, W.O.: Technological forecasting for decision making. Long Range Plan. **7**(1), 67–68 (1974). https://doi.org/10.1016/0024-6301(74)90087-9. Date of Access: 27 July 2023
56. Bass, F.M.: A new product growth for model consumer durables. Manag. Sci. **15**(5), 215–227 (1969). https://doi.org/10.1287/mnsc.15.5.215. Date of Access: 27 July 2023
57. Hall, B.: Innovation and diffusion. National Bureau of Economic Research, Inc, 2004. https://EconPapers.repec.org/RePEc:nbr:nberwo:10212. Date of Access: 27 July 2023
58. Romer, P.M.: Increasing returns and long-run growth. J. Polit. Econ. **94**(5), 1002–1037 (1986). https://doi.org/10.1086/261420. Date of Access: 27 July 2023
59. Stanton, B.F., Rogers, E.M.: Diffusion of innovations. J. Farm Econ. **45**(4), 898 (1963). https://doi.org/10.2307/1235767. Date of Access: 27 July 2023
60. Perez-Perez, C.: Toward a comprehensive theory of long waves. Long waves, depression, and innovation. Implication for national and regional economic policy: collaborative paper. IIASA, 103–107 (1985)
61. Denisov, V.: Dynamic models of cyclic development of photovoltaic power generation systems. Vidnovluvana energetika **1**(48), 36–43 (2017)
62. Saukh, S., Borysenko, A.: Mathematical Modeling of Electric Power Systems in Market Conditions: Monograph, 340 pp. "Tri K,", Kyiv (2020)
63. Denysov, V., Babak, V.: Software and information simulation complex of multi-node integrated and autonomous power and heat supply systems. Syst. Res. Energy. **2023**(3), 50–63 (2023). https://doi.org/10.15407/srenergy2023.03.050. Date of Access: 27 Sep 2023
64. Denysov, V. Software and information complex for district heat supply systems modeling. Syst. Res. Energy. **2022**(1), 38–45. https://doi.org/10.15407/srenergy2022.01.038. Date of Access: 27 Sep 2023
65. PLEXOS Energy Modeling Software. Energy Exemplar. https://www.energyexemplar.com/plexos. Date of Access: 21 July 2023

66. Shedding light on the future of the energy system. Shedding light on the future of the energy system. https://antares-simulator.org/. Date of Access: 21 July 2023
67. Flextool. IRENA – International Renewable Energy Agency. https://www.irena.org/Energy-Transition/Planning/Flextool. Date of Access: 21 July 2023
68. Denisov, V. The phased optimization concept of models for assessing and forecasting the structure and modes of use of energy complexes. In: Monografia Pokonferencyjna. Science, Research, Development: монографія. 19th ed, vol. 19, pp. 17–19. Warszawa, 2019
69. Ugolnitskii, G.A., Usov A.B. Control of complex ecological-economic systems. Autom. Remote. Control. **70**(5), 897–906 (2009). https://doi.org/10.1134/s0005117909050154. Date of Access: 27 July 2023
70. Denisov, V. Integrated Power System multi-node model, taking into account the nondispatchable of renewable energy sources. In: 2022 IEEE 8th International Conference on Energy Smart Systems (ESS), Kyiv, Ukraine, 12–14 October 2022. 2022. https://doi.org/10.1109/ess57819.2022.9969255. Date of Access: 27 July 2023
71. Kopp, R.E.: Pontryagin maximum principle. Math. Sci. Eng. **1962**, 255–279. https://doi.org/10.1016/s0076-5392(08)62095-0. Date of Access: 27 July 2023
72. Bellman, R., Dreyfus, S.: Functional approximations and dynamic programming. Math. Tables Other Aids Comput. **13**(68), p. 247 (1959). https://doi.org/10.2307/2002797. Date of Access: 27 July 2023
73. Denisov, V.A., Chuprina, L.V.: Technological conditions synchronous operation of united power system of Ukraine and neighboring ENTSO-E-member power systems. Probl. Gen. Energy. **2021**(3), 53–62 (2021). https://doi.org/10.15407/pge2021.03.053. Date of Access: 27 July 2023
74. On the Ratification of the Association Agreement between Ukraine, on the one hand, and the European Union, the European Atomic Energy Community, and their Member States, on the other hand: Law of Ukraine No. 1678-VII, dated September 16, 2014. https://zakon.rada.gov.ua/laws/show/1678-18#Text. Accessed 27 July 2023
75. On the Electricity Market: Law of Ukraine No. 2019-VIII, dated April 13, 2017. Updated as of July 27, 2023. https://zakon.rada.gov.ua/laws/show/2019-19#Text. Accessed 27 July 2023
76. Market Rules: Resolution No. 307 dated March 14, 2018
77. On Approval of the Market Rules for the Day-Ahead and Intraday Markets: Resolution No. 308 of the National Commission for State Regulation of Energy and Public Utilities dated March 14, 2018, as of December 2, 2022. https://zakon.rada.gov.ua/laws/show/v0308874-18#Text. Accessed 27 July 2023
78. Association Agreement between Ukraine, on the one hand, and the European Union, the European Atomic Energy Community, and their Member States, on the other hand: Agreement dated June 27, 2014, as of October 25, 2022. https://zakon.rada.gov.ua/laws/show/984_011#Text. Accessed 27 July 2023
79. Presentation by NPC Ukrenergo: Integration Plan of the Ukrainian Power System into ENTSO-E. Share and Discover Knowledge on SlideShare. https://www.slideshare.net/NPCUkrenergo/entsoe-233692950. Accessed 27 July 2023
80. World energy outlook 2019. OECD, 2019. https://doi.org/10.1787/caf32f3b-en. Date of Access: 27 July 2023
81. Denisov, V.A.: Determination of optimal operating modes of the Ukrainian power system when covering the daily schedule of electrical loads, ensuring the necessary volumes of redundancy and using storage capacities. Probl. Gen. Energy. **2020**(4), 33–44 (2020). https://doi.org/10.15407/pge2020.04.033. Date of Access: 27 July 2023
82. A comparative analysis of electricity generation costs from renewable, fossil fuel and nuclear sources in G20 countries for the period 2015–2030 / M. Ram et al. J. Clean. Prod. **199**, 687–704 (2018). https://doi.org/10.1016/j.jclepro.2018.07.159. Date of Access: 21 July 2023

83. Projected costs of generating electricity 2020—analysis. IEA. https://www.iea.org/reports/projected-costs-of-generating-electricity-2020. Date of Access: 21 July 2023
84. Timilsina, G.R.: Demystifying the Costs of Electricity Generation Technologies. Development Research Group, Development Economics.| The World Bank. https://documents1.worldbank.org/curated/en/125521593437517815/pdf/Demystifying-the-Costs-of-Electricity-Generation-Technologies.pdf. Date of Access: 21 July 2023
85. Non-baseload operation in nuclear power plants: load following and frequency control modes of flexible operation. International Atomic Energy Agency. https://www-pub.iaea.org/MTCD/Publications/PDF/P1756_web.pdf. Date of Access: 21 July 2023
86. Kostenko, G., Zaporozhets, A.: Transition from Electric Vehicles to Energy Storage: Review on Targeted Lithium-Ion Battery Diagnostics. Energies. **17**(20), 5132 (2024). https://doi.org/10.3390/en17205132
87. Parnell, G.S., et al.: Systems Engineering and Industrial Engineering. INCOSE Int. Symp. **32**(1), 1010–1021 (2022). https://doi.org/10.1002/iis2.12978. Date of Access: 21 July 2023
88. Calise, F., Figaj, R.: Recent Advances in Sustainable Energy and Environmental Development. Energies. **15**(18), 6534 (2022). https://doi.org/10.3390/en15186534. Date of Access: 21 July 2023
89. Denysov, V., et al.: Improving the model of long-term technological update of power system components. Syst. Res. Energy **2023**(2), 30–37 (2023). https://doi.org/10.15407/srenergy2023.02.030. Date of Access: 21 July 2023
90. Denysov, V., Kulyk, M., Babak, V., Zaporozhets, A., Kostenko, G.: Modeling nuclear-centric scenarios for Ukraine's low-carbon energy transition using diffusion and regression techniques. Energies. **17**(20), 5229 (2024). https://doi.org/10.3390/en17205229
91. Denysov, V.: Stochastic quasi-dynamic model of economic-technological impact of the life cycle of innovative technologies. In: Problems of Regulatory Support for Innovative Activity and Ways to Solve Them, pp. 35–37, Kyiv, September 27, 2017. Kyiv, 2017 (2017)
92. Lai, C.S., et al.: A review on long-term electrical power system modeling with energy storage. J. Clean. Prod. **280**, 124298 (2021). https://doi.org/10.1016/j.jclepro.2020.124298. Date of Access: 21 July 2023
93. Gaur, A.S., et al.: Long-term energy system planning considering short-term operational constraints. Energy Strat. Rev. **26**, 100383 (2019). https://doi.org/10.1016/j.esr.2019.100383. Date of Access: 21 July 2023
94. Izui, Y., Koyama, M.: Future energy and electric power systems and smart technologies. IEEJ Trans. Electr. Electron. Eng. **12**(4), 453–464 (2017). https://doi.org/10.1002/tee.22436. Date of Access: 21 July 2023
95. Report on the Assessment of the Adequacy of Generating Capacities to Meet Forecasted Electricity Demand and Ensure Necessary Reserves for 2022, approved by the NKREKP Resolution No. 664 dated April 11, 2023. NPC "Ukrenergo". Available at: https://ua.energy/zvit-z-otsinky-vidpovidnosti-dostatnosti-generuyuchyh-potuzhnostej/. Accessed: 21 July 2023
96. Maistrenko, N.Y.: The prediction of energy consumption for the sector of general government in the Ukrainian economy until 2040. Probl. Gen. Energy. **2022**(1–2), 82–89 (2022). https://doi.org/10.15407/pge2022.01-02.082. Date of Access: 02 Aug 2023
97. E-storage: Shifting from Cost to Value: World Energy Council (2016). http://www.worldenergy.org/publications/. Date of Access: 07 June 2023
98. Levelized cost of electricity and levelized avoided cost of electricity methodology supplement. Homepage—U.S. Energy Information Administration (EIA). https://www.eia.gov/renewable/workshop/gencosts/pdf/methodology_supplement.pdf. Date of Access: 21 July 2023
99. Levelized cost and levelized avoided cost of new generation resources in the annual energy outlook 2017. U.S. Energy Information Administration, 2017. 21 p.
100. Marín-Cano, C., Mejía-Giraldo, D.: Levelized avoided cost of electricity model based on power system operation. DYNA **85**(206), 79–84. September, 2018. Home Page. https://doi.org/10.15446/dyna.v85n206.69577. Date of Access: 27 Sep 2023

101. Diatlova, V., Petryk, I.: Modern approaches to the development of renewable energy as a component of Ukraine's innovative economy. Econ. Organ. Manag. **2**, 20–27 (2019). Available at: https://doi.org/10.31558/2307-2318.2019.2.2. Accessed: 1 Aug 2023
102. Kudria, S., et al.: Concept of the roadmap for the development of hydrogen energy in Ukraine until 2035. Renew. Energy **2019**(4), 22–28 (2019). Available at: https://doi.org/10.36296/1819-8058.2019.4. **59**, 22–28. Accessed: 1 Aug 2023
103. Bila, S.O.: Development of Renewable Energy as a Component of Transition to Sustainable Development: Experience of the EU and Ukraine [Thesis] (2018). Available at: http://er.nau.edu.ua/handle/NAU/40118. Accessed: 1 Aug 2023
104. Savytskyi, O., Aliieva, O.: Energy Transition—The biggest challenge of our time for Ukraine. Ukrainian Climate Network | Working Group of Non-Governmental Environmental Organizations on Climate Change (2023). Available at: https://ucn.org.ua/?p=4528. Accessed: 21 July 2023
105. Cabinet of Ministers of Ukraine: Approval of the Energy Strategy of Ukraine for the Period up to 2035 Security, Energy Efficiency, Competitiveness: Order No. 605-p (2017), August 18, 2017. Available at: https://zakon.rada.gov.ua/laws/show/605-2017-p#Text. Accessed: 7 Aug 2023
106. Denisiuk, M.: What will the energy sector look like after the war. Economic Truth (2023). Available at: https://www.epravda.com.ua/columns/2023/07/4/701854/. Accessed: 21 July 2023

Chapter 2
Mathematical Models and Programming Tools for Optimizing the Composition and Operating Modes of Energy Systems Under Rapid Growth of Renewable Energy Capacities

2.1 Dynamic Model for Cyclic Development of Generation Systems

The fundamental model [1] of innovation technology k diffusion, modified for generation system technologies, is represented by formula (2.1):

$$\frac{dP^k(\tau)}{d\tau} = g(\tau)\left(P^{\Sigma k}(\tau) - P^k(\tau)\right), \tag{2.1}$$

where

- $g(\tau)$ is the diffusion coefficient;
- $P^k(\tau)$ is the current capacity of the technology;
- $P^{\Sigma k}(\tau)$ is the maximum possible capacity of the technology.

Depending on the formula for the diffusion coefficient $g(\tau)$, three models of innovation diffusion are proposed: the external influence model, the internal influence model, and Frank Bass's mixed influence model (formula 1.1 in Chap. 1), which combines both previous models and is most commonly used in research.

The main difference of the proposed in this work dynamic model of cyclic development of generation systems is the explicit consideration of the influence of economic and technological indicators of the national economy and production development, which are represented in the form of quasi-dynamic functions with discrete stochastic variables (Fig. 2.1). These indicators, which are components of the proposed functional of economic-technological impact, incorporated into the new generalized economic-mathematical model of technology development in the form of a sum of several logistic curves, are calculated using the modified of the Bass mixed influence model formula (2.2):

A. Zaporozhets et al., *Structure Optimization of Power Systems with Renewable Energy Sources*, Studies in Systems, Decision and Control 583,
https://doi.org/10.1007/978-3-031-83697-8_2

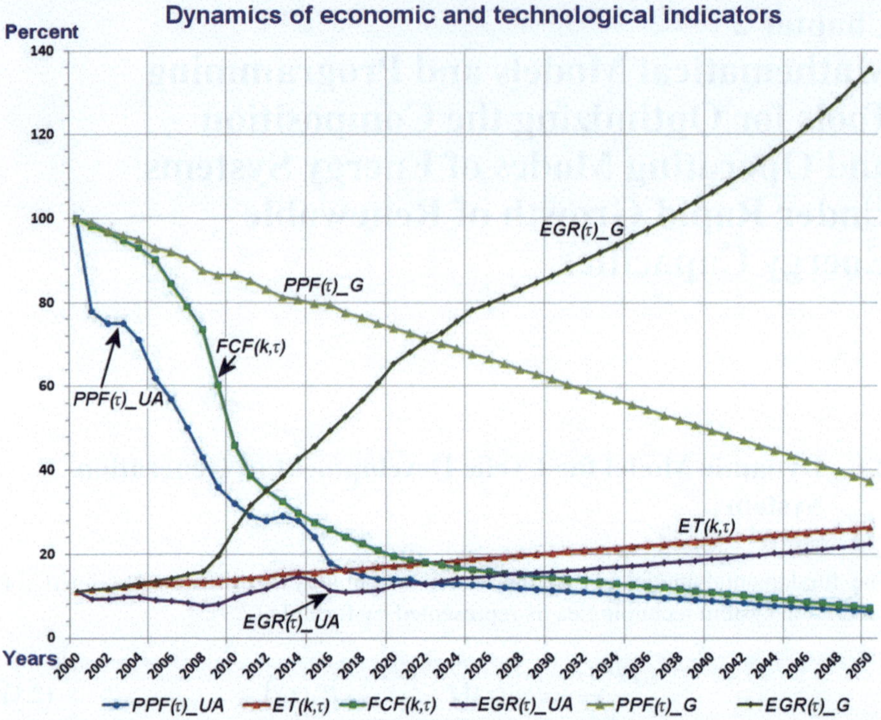

Fig. 2.1 Dynamics of economic and technological indicators of economic and production development for Ukraine (UA) and Germany (G): $PPF(\tau)$—regional purchasing power coefficient., $ET(k, \tau)$—technology efficiency, $FCF(k, \tau)$—coefficient of final cost, $EGR(\tau)$—economic growth rate

$$\frac{dP^k(\tau)}{d\tau} = K\left\{p + \frac{q}{P^{\sum k}}P^k(\tau)\right\}\left(P^{\sum k}(\tau) - P^k(\tau)\right). \tag{2.2}$$

The parameters K, p, q are determined based on the approximation of their known actual (up to 2022) values using the least squares method.

To account for the indicated indicators, the formula for the diffusion coefficient in the mixed influence model of Bass [2], which has the form $g(\tau) = p + qN(\tau)$, takes the following form:

$$g(\tau) = K_m F[PPF(\tau), ET(k, \tau), FCF(k, \tau), EGR(\tau)]\left\{p + \frac{q}{M(\tau)}N(\tau)\right\},$$

where

- K_m—scaling coefficient;
- $F[PPF(\tau), ET(k, \tau), FCF(k, \tau), EGR(\tau)]$—the functional of economic-technological influence, which depends on economic and technological indicators of economic and production development.

The following formula was used for calculating the functional:

$$F[PPF(\tau), ET(k,\tau), FCF(k,\tau), EGR(\tau)] =$$
$$F(\tau) = \frac{(PPF(\tau)/PPF(\tau_0)) \cdot ET(k,\tau) \cdot EGR(\tau)}{FCF(k,\tau)/FCF(\tau_0)},$$

where τ_0—the initial moment, for example, the year 2000. Thus, the functional $F(\tau)$ is a dimensionless quantity. The components included in it, $PPF(\tau)/PPF(\tau_0)$ and $FCF(\tau)/FCF(\tau_0)$, are denoted as follows:

- $KPPF(\tau) = PPF(\tau)/PPF(\tau_0)$—the adjusted purchasing power parity coefficient of the region;
- $KF(k, \tau) = FCF(k, \tau)/FCF(k, \tau_0)$—the adjusted coefficient of average final cost for the consumer.

Quasidynamic functions with discrete stochastic variables that enter the functional $F(\tau)$ are called coefficients of economic and technological influence and are expressed in percentages. In this case, the values are equal to 100% at the moment τ_0. Therefore:

$$F(\tau) = \frac{KPPF(\tau) \cdot ET(k,\tau) \cdot EGR(\tau)}{KF(k,\tau)}. \quad (2.3)$$

With the considerations provided, the equation for the model of economic-technological influence on the development of energy technology k takes the following form:

$$\frac{dP^k(\tau)}{d\tau} = F(\tau)\left(P^{\Sigma k}(\tau) - P^k(\tau)\right),$$

where

- $P^k(\tau)$—cumulative power of technology utilization k at the moment τ;
- $P^{\Sigma k}(\tau)$—maximum achievable capacity of the technology k;
- p—coefficient of innovation, or the intensity of influence from the external market environment where a certain innovation is already being used;
- q—coefficient of information transmission about innovation among participants.

In turn, the total installed capacity of the technology at moment $\tau + 1$ is determined by the following equation:

$$P^k(\tau+1) = P^k(\tau) + K_m F(\tau)\left\{p + \frac{q}{P^{\Sigma k}(\tau)} P^k(\tau)\right\}\left(P^{\Sigma k}(\tau) - P^k(\tau)\right).$$

This model explicitly takes into account the influence of economic and technological factors and reflects the development of technology in the form of a logistic curve. To represent the life cycle with two or more peaks, a model in the form of a generalized logistic curve, which is the sum of several logistic curves, is proposed:

$$P^k(\tau+1) = P^k(\tau) + K_m \sum_{j=1}^{J} F(\tau)\left\{p_j + \frac{q_j}{P_j^{\sum k}(\tau)} P_j^k(\tau)\right\}\left(P_j^{\sum k}(\tau) - P_j^k(\tau)\right), \tag{2.4}$$

where j K_{KD}(t)—number of logistic curves, the total quantity of which is equal to J.

In general, both J and the parameters p_j, q_j, $P^{\sum k}{}_j(\tau)$ can vary for each j individually, but it is assumed that p_j and q_j remain constant throughout the modeling horizon. The results of the calculations of the forecast parameters of installed solar energy capacity for Ukraine and Germany until 2050 are presented in Chap. 4.

The *dynamic model of cyclical development* of power systems has allowed for the development of a new *method for economic-technological forecasting of optimal functioning parameters and the development of the power system*. It has also led to a new *model for long-term technological upgrading of consumption structure and generating capacity* of the energy system in Ukraine.

2.2 Method of Economic-Technological Forecasting of Optimal Functioning Parameters and Development of the Energy System, Along with the Model for Long-Term Technological Upgrading of Consumption Structure and Generating Capacity of the Energy System in Ukraine

The method of economic-technological forecasting of optimal functioning parameters and development of the energy system, along with the model for long-term technological upgrading of consumption structure and generating capacity of the energy system in Ukraine. [3], which are the components of this method designed for investigating short- and long-term scenarios. The model is a specific case of a hierarchically controlled quasi-dynamic system [4] and a quasi-dynamic model of economic-technological influence [5] on the life cycle of innovative technologies. The main difference of the proposed model is the explicit consideration of the impact of economic and technological development indicators of the national economy and production, represented in the form of quasi-dynamic functions with discrete stochastic variables. Actual and forecasted values of these variables at the corresponding time step of the modeling horizon were used in the calculations. Model [3] is aimed at researching directions and optimal parameters for technological upgrading of components in the energy system from a long-term perspective.

The most widely used productivity factors in modeling scenarios for the development of energy systems are taken into account.

Levelized Cost of Electricity (LCOE)

Currently, in many cases, the assessment of decision-making effectiveness is based on calculating the levelized cost of electricity (LCOE). LCOE is often used as a convenient comprehensive indicator of the overall competitiveness of different generating technologies. This indicator represents the cost per kilowatt-hour (taking into account the discounted value of currency) for the construction and operation of an energy facility throughout its anticipated life cycle, including decommissioning costs.

The main input data for the calculation include capital costs, fuel costs, fixed and variable operational costs, maintenance costs, financing costs, and the anticipated utilization factor for each type of generating facility. The weight of these factors varies depending on the technology. For technologies based on solar energy or wind energy, which have no fuel costs and relatively low variable costs, LCOE is approximately proportional to the capital cost of generating capacities. For technology with significant fuel costs, fuel and operational costs significantly influence LCOE.

The availability of various incentives, including state or federal tax credits, can also affect the calculation of LCOE. Like in any design project, there is uncertainty about all these factors—their values may vary regionally and over time as technology develops and fuel prices change. It's important to note that actual investment costs depend on specific technological and regional project characteristics, involving many other factors.

Levelized Avoided Cost of Electricity (LACE) [6–8] as an additional indicator.

Since forecasted utilization rates, existing resource sets, and required capacity volumes can vary significantly in different regions where new generation capacities may be needed, direct comparison of technologies based on LCOE as a method for assessing the economic competitiveness of different generating projects can often be problematic and misleading. Conceptually [6–8], a better assessment of economic competitiveness can be achieved by considering avoided costs, namely, the cost of generating with resources already available, compared to the electricity the new-generation project is expected to generate, as well as its LCOE. The volume of avoided costs, which provides a measure of confidence for the annual economic value of the candidate project, can be summed up over its financial life cycle and divided by the average annual generation volume of the project. This allows for the determination of the specific avoided cost of electricity (LACE).

Subsequently, for the candidate project, the LACE value can be compared to its LCOE value to determine whether the benefits obtained from the project exceed its cost. When considering multiple competing technologies, comparing each project's LACE to its LCOE can be used to select the most efficient project. Evaluating LACE is more complex than evaluating LCOE because it requires information on how the system could operate without modification. Calculating LACE is based on forecasting the marginal cost of energy and capacity using either existing or developing technology by a certain future date, allowing the investor to assess the potential value of project implementation. An example of a possible formula for calculating LACE is provided in [8].

Long-term forecasts often use indicators like LACE and LCOE, which are typically more representative than using just LCOE alone. However, it's important to note that LACE and LCOE estimates are simplifications of modeled decisions and cannot fully account for all factors for decision-making or comparing modeling results.

Calculations based on LACE and LCOE estimates, which are typically conducted using forecasted sets of capital and operational costs, are subject to uncertainty due to projected fuel prices and potential regulatory actions. This uncertainty should be taken into account when analyzing technology choices in the power sector based on calculated LCOE or LACE.

Estimates of LCOE for various prospective generating technologies planned for commissioning after 2022 and in 2040 are provided in [6–9].

The model of long-term technological renewal of the structure of consumption and generating capacity of the IPS of Ukraine [3] presented as a hierarchy of scenarios. At the top level of the hierarchy, aggregated energy supply technologies are involved. The system state matrix $\Omega(k, \tau)$ reflects the structure of energy generation, supply and consumption volumes at the step τ, $\tau = 0, 1, 2, \ldots, T$ of the modeling horizon. All the technologies used $k = 1,\ldots,K$ are involved in ensuring the balance of energy supplied $E_{S\tau}^{\Sigma}$ and consumed $E_{C\tau}^{\Sigma} = \sum_{k=1}^{K} E_{k\tau}^{Consum}$ at each step τ. The main constraints of the model are to maintain a balance of total power P_{τ}^{Σ}, maneuvering power $P_{m\tau}^{\Sigma}$ and volumes of generated $E_{G\tau}^{\Sigma} = \sum_{k=1}^{K} E_{G\tau}^{k}$, supplied $E_{S\tau}^{\Sigma}$ and consumed $E_{C\tau}^{\Sigma}$ energy, provided that all parameters belong to a set of possible states. A measure μ_τ of inconsistency between the vectors of supplied and consumed energy has been introduced. The initial information for modeling is:

- target sequence of annual energy consumption $E_{C\tau}^{\Sigma} = \sum_{k=1}^{K} E_{k\tau}^{Consum}$;
- initial state of the total supply vector E_{S0}^{Σ};
- Net Benefit vector $NB_k = L_k(LACE_k - LCOE_k)$—integral differences in parameters for each of the aggregated technologies involved in the calculation;
- functional of economic and technological influence $F(k, \tau) = \frac{KPPF(\tau) \cdot ET(k,\tau) \cdot EGR(\tau)}{KFCF(k,\tau)}$;
- projected cost of technology components: $C_{k\tau}^{cap}$—investment, $C_{k\tau}^{FO}$—fixed and $C_{k\tau}^{VO}$—variable operating costs.

A set of admissible states $\Phi_{k\tau}\{NB_k, F(k, \tau)\}$ for the trajectory development of the aggregated technology k is calculating. The problem of calculating such a supply vector $E_{S\tau}^{\Sigma}(\Phi_{k\tau})$ that minimizes the measure μ^{Σ} of the total inconsistency of the vectors of supplied and consumed energy is solved:

$$\mu^{\Sigma} = \sum_{\tau=1}^{\mathrm{T}} \left(E_{S\tau}^{\Sigma} - E_{C\tau}^{\Sigma} \right) \forall \;\; u(\tau, k), \xi(\tau, k)) \rightarrow \min, \tag{2.5}$$

where, at each step $\tau = 0, 1, 2, \ldots, T$, for each aggregated technology k: $E_{S\tau}^{\Sigma}$, $E_{C\tau}^{\Sigma}$—total volumes of energy supplied and consumed; $u(\tau, k)$—vector of control actions;

$\xi(\tau, k)$—vector of random external perturbations. The optimal target trajectory of the supply vector $E_{S\tau}^{\sum\_OPT}(\Phi_{k\tau})$ determined on the basis of forecasts the economy general development and related the consumption structure $E_{C\tau}^{\sum} = \sum_{k=1}^{K} E_{C\tau}^{k}$.

The values of the total supply volumes obtained at the upper level and the values of the vector components $E_{S\tau}^{\sum} = \sum_{k=1}^{K} E_{S\tau}^{k}$ for each of the aggregated technologies are the initial data for calculations: the required balance $P_{\tau}^{\sum}$ of total power, the required structure $P_{m\tau}^{\sum}$ of maneuvering capacities, the volume of total and component-by-component generation $E_{G\tau}^{\sum} = \sum_{k=1}^{K} E_{G\tau}^{k}$, the volume of required investments $C_{k\tau}^{\sum cap} = \sum_{k=1}^{K} C_{k\tau}^{cap}$, the volume of fixed operating costs $C_{k\tau}^{\sum co} = \sum_{k=1}^{K} C_{k\tau}^{co}$, the amount of variable operating costs $C_{k\tau}^{\sum vo} = \sum_{k=1}^{K} C_{k\tau}^{vo}$.

At the next levels of modeling, similar calculations are performed within each of the aggregated technologies, i.e. the components of the scenarios for the development of the corresponding vectors of each of the aggregated technologies are calculated.

Using this model, a forecast of the consumption and generation structure of electricity in the power system of Ukraine until 2040 has been calculated. According to the method of economic-technological forecasting of parameters for optimal functioning and development of the energy system, the forecast is calculated in three stages:

1. In the first stage, using the generalized logistic model (2.4) and (2.6), pessimistic and optimistic forecasts of electricity consumption volumes in the power system of Ukraine until 2040 have been calculated (Fig. 2.2).

$$E_{C}^{\sum}(\tau+1) = E_{C}^{\sum}(\tau) + K_m \sum_{j=1}^{J} F(\tau)\left\{p_j + \frac{q_j}{E_{Cj}^{\sum}(\tau)} E_{Cj}(\tau)\right\}\left(E_{Cj}^{\sum}(\tau) - E_{Cj}(\tau)\right). \tag{2.6}$$

2. In the second stage, using the modified formula of net benefit $NB_k = K_m(LACE_k - LCOE_k)$ [8] and the formula to minimize the aggregate cost of generation $C_{S\tau}^{\sum} = \sum_{k=1}^{K} NB_k \cdot E_{S\tau}^{k} \to \min$, the ratios of the required capacity and volume of total annual generation of aggregated technologies are calculated.
3. At the third stage the problem of minimization of the inconsistency of the total volume of generation and consumption of electric power (2.5) was solved (Figs. 2.3 and 2.4).

More detailed input data and the results of the forecast calculations of the structure of electricity consumption and generation of the Integrated Power System (IPS) of Ukraine up to 2040 for each stage are presented in Chap. 4.

Thus:

1. Using a new mathematical model of innovation diffusion in the energy sector, scenarios for the development of solar power plants (SPPs) in Germany and Ukraine were calculated (presented in Chap. 4).

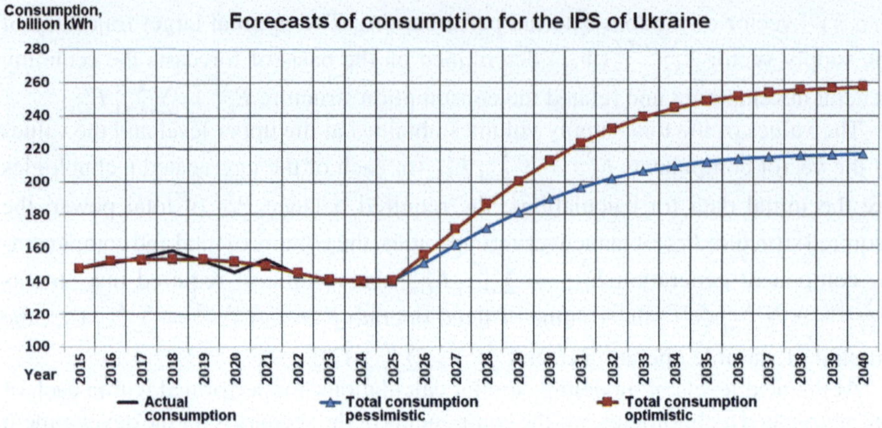

Fig. 2.2 Pessimistic and optimistic forecasts of electricity consumption volumes in the power system of Ukraine until 2040

Fig. 2.3 The structures development scenarios of the volumes of generation for the IPS of Ukraine until 2040

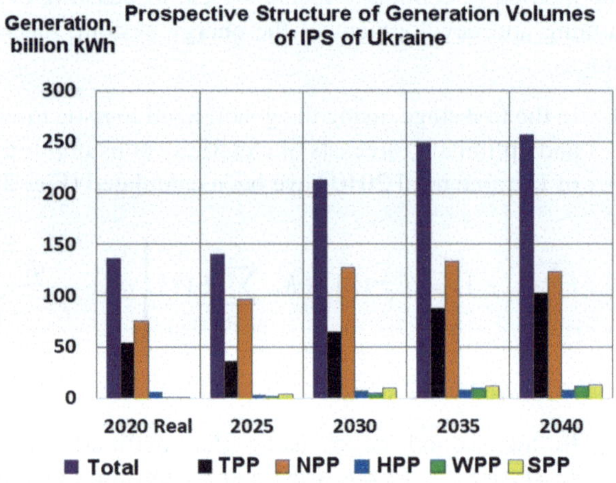

2. The method of economic-technological forecasting of the parameters of optimal functioning and development of the energy system, along with the model of long-term technological renewal of the energy system structure, allowed the development of forecasts for the structure of electricity consumption and generation of the Integrated Power System (IPS) of Ukraine up to 2040, enabling the assessment of the potential for improving the components of the IPS of Ukraine.

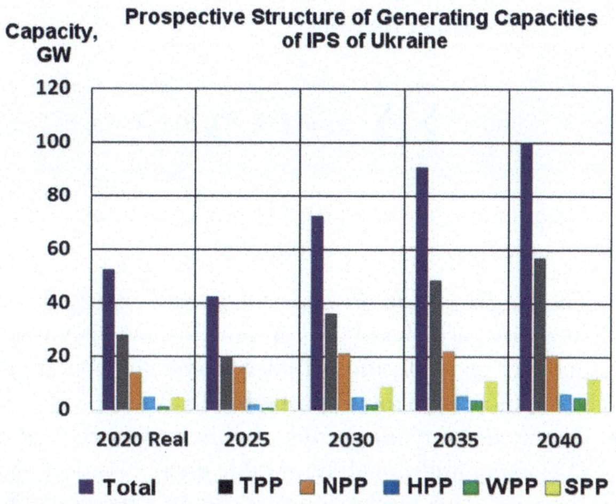

Fig. 2.4 The structures development scenarios of the generating capacities for the IPS of Ukraine until 2040

2.3 Model of a Hierarchically Managed Quasi-Dynamic Energy System

Based on the considerations outlined in Sect. 1.6, a model of a hierarchically managed quasi-dynamic energy system (HQDS) with $r \in R$ levels of administrative-territorial hierarchy and sectoral (subsectoral) infrastructure, detailed by the structure of its technological $k \in K$ content, has been developed [10, 11].

The task of hierarchical management of such a dynamic system is formulated as follows. Over the modeling horizon T, for all: $\tau = 1, 2, \ldots, T$; $r = 1, 2, \ldots R$; $k = 1, 2, \ldots K$:

- $\Omega(\tau,r,k)$—the state vector of the system,
- $\Phi(\tau,r,k)$—the set of allowable states of the system,
- g—the economic-technological impact functional, which includes:
  - $\omega(\tau, r, k)$—elements of the state matrix,
  - $u(\tau, r, k)$—elements of the control actions matrix,
  - $\xi(\tau, r, k)$—random elements of the external impact matrix, for example, in our case, the generating capacities of renewable energy sources (RES),
- μ—the optimality criterion,
- $U(\tau, r, k)$, $\Xi(\tau, r, k)$—sets of possible values of control actions and random external impacts.

When the following conditions are met: $\Omega(\tau, r, k) \in \Phi(\tau, r, k)$, $u(\tau, r, k) \in U(\tau, r, k)$, $\xi(\tau, r, k) \in \Xi(\tau, r, k)$, $\omega(\tau, r, k) \in \Omega(\tau, r, k)$, under the influence of $u(\tau, r, k)$, $\xi(\tau, r, k)$, the system transitions to the next state:

$$\Omega(\tau, r, k)|u(\tau, r, k), \xi(\tau, r, k) \Rightarrow \Omega(\tau + 1, r, k).$$

The optimality criterion:

$$\mu_\tau = \sum_r^R \sum_k^K g(\omega(\tau, r, k), u(\tau, r, k), \xi(\tau, r, k)) \rightarrow \min / \max .$$

Specific cases of the [10, 11] model of a hierarchically managed quasi-dynamic system are the following two models:

- *The model of a multi-node integrated energy system* [10, 11] with integer optimization variables of the composition and operating modes of generating, reserve, and storage capacities of energy systems operating within a single synchronous zone, the continental Europe ENTSO-E.
- *The model for minimizing the average weighted cost of electricity production* [12] using traditional, renewable energy sources (RES), and energy storage technologies. This model allows evaluating the potential for possible reduction in the average hourly cost of electricity production and based on forecast load profiles "one day ahead," solve the problem of forecasting the selection of optimal operating modes for the use of generating and storage capacities of the system. It also enables hourly minimization of the cost of electricity production.

2.4 Model of a Multi-node Integrated Energy System

The model of a multi-node integrated energy system [10] is a specific case of the model of a hierarchically managed quasi-dynamic system [4]—a mathematical model with integer optimization variables for the composition and operating modes of generating, reserve, and storage capacities of energy systems operating within a single synchronous zone, the continental Europe ENTSO-E. The model belongs to the class of optimal generation scheduling problems with the criterion of minimizing production costs and import/export of electricity. In its most general form, the model can be described as follows:

$$\sum_r^R \sum_\tau^T \left\{ \sum_g^G \left(P_{r\tau g}^{Gen} c_{r\tau g}^{Gen} + S_{r\tau g}^{Start} c_{r\tau g}^{Start} \right) + \sum_k^R P_{rk\tau}^{Imp} c_{k\tau}^{Imp} (1 - a_{kr}^{Transf}) \right\} \rightarrow min,$$

Production—consumption—import—export balance:

$$\sum_g^G P_{r\tau g}^{Gen} + \sum_k^R \left(P_{rk\tau}^{Imp} - P_{rk\tau}^{Exp} \right) = D_{r\tau}; \forall \tau \in T; \forall r \in R.$$

Power flow balance:

Constraints on transmission capacity (Net Transfer Capacity):

$$P_{rk\tau}^{Exp} \le F_{kr}^{NTC}; P_{rk\tau}^{Imp} \le F_{kr}^{NTC}; \forall\tau \in T; \forall r \in R; \forall k \in R(k = r).$$

Possible constraint on import balance—the import from node r to node k must equal the export from node k to node r:

$$P_{rk\tau}^{Imp} - P_{rk\tau}^{Exp} = 0; \forall\tau \in T; \forall r \in R; \forall k \in R(k = r).$$

Here: T—period (time in hours); $R(K)$—set of nodes; G—set of generating capacities; P—power; c—cost; α—loss coefficient; D—consumption; F—maximum Net Transfer Capacity.

The initial information for assessing and forecasting the optimization of the composition and operating modes of generating, reserve, and storage capacities of energy systems operating within the single synchronous zone of continental Europe ENTSO-E is a set of vectors and matrices of output parameters for each participating country for each time step $\tau = 1, 2, \ldots, T$ on the modeling horizon:

- S—list of countries s;
- J—list of technologies, $j = 1, \ldots, J$;
- $L(\tau, s)$—forecasted or calculated by the dispatch service of each country, the matrix of hourly load profile;
- $C(c, j, s)$—matrix of technology costs;
- $C^{\mathrm{FUEL}}(f, j, s)$—fuel price matrix f;
- $W^{CO_2}(j, s)$—emissions matrix CO_2;
- $P^{\mathrm{RES}}(j, s, \tau)$—hourly power output of renewable energy technologies (RETs);
- $(P^{inst}(j, s), P^{\max}(j, s), P^{\min}(j, s))$—matrix of parameters and constraints for non-dispatchable renewable energy sources.
- $\theta(\mathrm{n(j,s)}, P^{\mathrm{inst}}(\mathrm{j,s}), P^{\max}(\mathrm{j,s}), P^{min}(\mathrm{j,s}), P^{\mathrm{ramp}}(\mathrm{j,s}))$—matrix of parameters and constraints for dispatchable technologies in which:
  - $n(j, s)$—number of power units for technology j in the country s;
  - $P^{inst}(j, s), P^{\max}(j, s), P^{\min}(j, s)$—installed, maximum, and minimum permissible power;
  - $E^{\max}(j, s)$—maximum electricity production during the day;
  - $P^{ramp}(j, s)$—permissible rate of power change.

All the mentioned resources are used to solve a problem where at each time step τ of the modeling horizon T, the task of finding the optimal set of generation, reserve, storage, import/export modes is solved while adhering to technological constraints. The solution to the problem is a set of matrices that provide optimal conditions for: generation $G(L, C, P^{RES}, \Theta, \Psi)$, reserve $R(L, C, P^{RES}, \Theta, \Psi)$, storage $A(L, C, P^{RES}, \Theta, \Psi)$ and import/export (volume of cross-border flows) $F(L, C, P^{RES}, \Theta, \Psi)$.

When forming the input data parameters, the requirements specified in the Transmission System Code [13] are taken into account to ensure the following:

1. Frequency restoration.
2. Reserving and supporting operation in emergency and normal modes.
3. Available volumes of cross-border technological electricity flows while operating within the single synchronous zone of continental Europe ENTSO-E energy systems of Ukraine, Poland, Hungary, and Romania.

In the work [11], a simplified version of the mathematical model is presented, which is built for the IPS of Ukraine only. The main sets, parameters, and variables of the model constructed for the energy system of Ukraine are provided in Appendix B.

Works [14–16] are dedicated to the analysis of actual annual load coverage schedules of the power system of Ukraine and the development of new dispatching modes for hydro units of hydroelectric power plants (HPPs) [16, 17]. The results of this analysis justify the development of mathematical models, the implementation of which will allow calculating a necessary number of dispatching mode options for hydro units of HPPs within an acceptable time frame to ensure the balance of the power system. Specifically, this can be achieved by introducing efficient utilization modes of units that provide reserve and energy storage. In case of insufficiency of these measures, additional, significantly more maneuverable energy storage and reserve power generating capacities can be introduced.

The simplified version of the mathematical model [11], built for the energy system of Ukraine, unlike the models proposed in [14, 15], allows for the use of large-capacity storage units and enables the investigation of the effectiveness of their combined utilization.

Below are the main formulas and constraints of the optimization model.

In the modified mathematical model, the optimality criterion is the sum of electricity production costs over the day (2.7), considering the generating capacities involved in covering the load profile, taking into account the costs of operating hydro units of pumped storage power plant (PSPP) in pumping mode and the costs of operating the storage unit in charge mode.

$$\sum_{t}^{T}\left[\begin{array}{l} C_{NPP}^{G}\cdot P_t^{NPP}+C_{PV}^{G}\cdot P_t^{PV}+C_{WIND}^{G}\cdot P_t^{WIND}+C_{WPP}^{G}\cdot P_t^{WPP}+\sum\limits_{tpp}^{TPP}C_{TPP}^{G}\cdot P_t^{TPP}\\ +\sum\limits_{hp}^{HP}\left(C_{HP}^{G}\cdot G_t^{HP}+C_{HP}^{P}\cdot P_t^{HP}\right)+C_{B}^{G}\cdot P_t^{BG}+C_{B}^{P}\cdot P_t^{BC}+C_{I}^{G}\cdot P_t^{I}\\ +C_{E}^{G}\cdot P_t^{E}+C_{deficit}^{P}\cdot P_t^{P\_deficit}+C_{overplus}^{P}\cdot P_t^{P\_overplus} \end{array}\right]$$

$$\to \min. \tag{2.7}$$

Additional "penalty" costs are considered in case it is impossible to maintain balance (2.8) without using penalty variables $P_t^{P\_deficit}$ and $P_t^{P\_overplus}$.

$$\sum_{t}^{T}\left[\begin{array}{l} P_t^{NPP} + P_t^{PV} + P_t^{WIND} + P_t^{WPP} + \sum_{tpp}^{TPP} P_t^{TPP} + \sum_{hp}^{HP} \left(G_t^{HP} - P_t^{HP}\right) \\ + P_t^{BG} - P_t^{BC} + P_t^{I} - P_t^{E} + P_t^{P\_deficit} - P_t^{P\_overplus} \end{array}\right] = D_t. \tag{2.8}$$

Understood, the need to use these penalty variables indicates a mismatch between the parameter sets and components of the power system with the specified load profile, requiring an analysis of the reasons for this situation and the search for ways to resolve the identified mismatch.

For each moment in time, balance (2.8) between consumption and electricity production must be maintained, taking into account the possibility of using a portion of generated energy for operating hydro units of pumped storage power plant (PSPP) in pumping mode, for energy storage in charge mode of the battery storage, for electricity export, and with the ability to ensure balance through imports and penalty variables.

During the modeling process, technological and algorithmic constraints are used, allowing for a significant refinement of modeling results, forecasting the limits of acceptability of individual parameters, and the overall applicability of the model. The most significant constraints are listed below.

The total power output supplied to the grid by all nuclear power plants (NPP) is specified:

$$P_t^{NPP} = P_{NPP}^{\Sigma}. \tag{2.9}$$

The constraint of allowable capacity for SPPs and wind power plants (WPP) within a specific time period:

$$P_t^{PV} = \left[\begin{array}{l} P_t^{PV} : P_t^{PV} \le P_{PV}^{\sum G} \\ P_{PV}^{\sum G} : P_t^{PV} > P_{PV}^{\sum G} \end{array}\right.; \forall t \in T; \tag{2.10}$$

$$P_t^{WIND} = \left[\begin{array}{l} P_t^{WIND} : P_t^{WIND} \le P_{WIND}^{\sum G} \\ P_{WIND}^{\sum G} : P_t^{WIND} > P_{WIND}^{\sum G} \end{array}\right.; \forall t \in T. \tag{2.11}$$

The most significant operational constraints of pumped storage power plant (PSPP).

The amount of energy stored by each pumped storage power plant (PSPP) at a specific moment t:

$$E_t^{HP} = \left[\begin{array}{l} 0 : t \le 1 \\ \sum_{1}^{t-1} \left(P_t^{HP} - G_t^{HP}\right) \cdot G_{HP} : t > 1 \end{array}\right.; \forall t \in T. \tag{2.12}$$

The total amount of energy stored by all pumped storage power plants (PSPPs) at a specific moment t:

$$E_t^{HP\_\Sigma} = \left[\sum_{hp}^{HP} \sum_{1}^{t} E_t^{HP}; \forall hp \in HP; \forall t \in T. \right. \tag{2.13}$$

The constraint on the hourly consumption volume of PSPPs stored energy:

$$G_t^{HP} \cdot G_{HP} \le E_t^{HP}; \forall hp \in HP; \forall t \in T. \tag{2.14}$$

Each pumped storage power plant (PSPP) must utilize all the stored water:

$$\sum_{t}^{T} G_t^{HP} \cdot G_{HP} = \sum_{t}^{T} P_t^{HP} \cdot G_{HP}; \forall hp \in HP; \forall t \in T. \tag{2.15}$$

The constraint on the pumping capacity in pumping mode:

$$\sum_{hp}^{HP} P_t^{HP} \le P_{HP}^{\Sigma P}; \forall hp \in HP; \forall t \in T. \tag{2.16}$$

The constraint on the power output in generation mode:

$$\sum_{hp}^{HP} G_t^{HP} \le P_{HP}^{\Sigma G}; \forall hp \in HP; \forall t \in T. \tag{2.17}$$

Prohibition of simultaneous operation of a hydro unit in pumping mode and generation mode:

$$Y_{bt}^{HP\_P} + Y_{bt}^{HP\_G} \le 1; \forall hp \in HP; \forall t \in T. \tag{2.18}$$

The use of pumped storage power plants (PSPPs) to provide reserves for load balancing and system stabilization for each modeling period. The maximum and minimum hourly power reserve capacity of PSPPs for load balancing and system stabilization are technological constraints.

Maximum hourly power reserve capacity of PSPPs for load balancing:

$$R_t^{HP\_UP} = \left[\begin{array}{l} R_{HP}^{SWITCH} \cdot E_t^{HP\_\Sigma} : R_{HP}^{SWITCH} \cdot E_t^{HP\_\Sigma} \le R_{HP}^{MAX\_UP} \\ R_{HP}^{MAX\_UP} : R_{HP}^{SWITCH} \cdot E_t^{HP\_\Sigma} > R_{HP}^{MAX\_UP} \end{array} \right. ; \forall hp \in HP; \forall t \in T. \tag{2.19}$$

The minimum hourly power reserve capacity of PSPPs for unloading:

$$R_t^{HP\_DOWN} = \begin{cases} R_{HP}^{SWITCH} \cdot \sum_{hp}^{HP} \left(P_t^{HP} + G_{HP} \cdot G_t^{HP}\right) : R_{HP}^{SWITCH} \cdot \sum_{hp}^{HP} \left(P_t^{HP} + G_{HP} \cdot G_t^{HP}\right) \\ \geq R_{HP}^{MIN\_DOWN} R_{HP}^{MIN\_DOWN} : \quad R_{HP}^{SWITCH} \cdot \sum_{hp}^{HP} \left(P_t^{HP} + G_{HP} \cdot G_t^{HP}\right); \\ < R_{HP}^{MIN\_DOWN} \\ \forall hp \in HP; \forall t \in T. \end{cases} \tag{2.20}$$

The most significant operational constraints of Thermal Power Plants (TPPs). Constraint on the maximum number of starts of a TPP unit:

$$\sum_t^T Y_{bt}^{TPP} \leq k_{TPP}^{GD}; \forall t \in T. \tag{2.21}$$

Constraint on the minimum composition of TPP units in covering the electricity load curve throughout the day:

$$\sum_t^T Y_{ot}^{TPP} \leq k_{TPP}^{G\_MIN}; \forall t \in T. \tag{2.22}$$

Constraint on the power output of each TPPs unit during a specific period of time:

$$P_t^{TPP} \geq Y_{ot}^{TPP} \cdot P_{TPP}^{MIN}; \forall tpp \in TPP; \forall t \in T; \tag{2.23}$$

$$P_t^{TPP} \leq Y_{ot}^{TPP} \cdot P_{TPP}^{MAX}; \forall tpp \in TPP; \forall t \in T. \tag{2.24}$$

Constraint on the minimum and maximum change in power output of a TPP unit compared to the previous period:

$$P_t^{TPP} - P_{t-1}^{TPP} \leq Y_{ot}^{TPP} \cdot P_{TPP}^{RUMP\_MAX}; \forall tpp \in TPP; \forall t \in T; \tag{2.25}$$

$$P_t^{TPP} - P_{t-1}^{TPP} \geq Y_{ot}^{TPP} \cdot P_{TPP}^{RUMP\_MIN}; \forall tpp \in TPP; \forall t \in T. \tag{2.26}$$

The use of TPPs to provide reserves for load balancing and system stabilization for each modeling period. The maximum and minimum hourly power reserve capacity of TPPs for load balancing and system stabilization are technological constraints.

Maximum hourly power reserve capacity of TPPs for load balancing:

$$R_t^{TPP\_UP} = \begin{cases} R_{TPP}^{SWITCH} \cdot \sum_{tpp}^{TPP} (Y_{ot}^{TPP} \cdot P_{TPP}^{MAX} - P_t^{TPP}) : R_{TPP}^{SWITCH} \cdot \sum_{tpp}^{TPP} (Y_{ot}^{TPP} \cdot P_{TPP}^{MAX} - P_t^{TPP}) \le R_{TPP}^{UP} \\ R_{TPP}^{UP} : \qquad R_{TPP}^{SWITCH} \cdot \sum_{tpp}^{TPP} (Y_{ot}^{TPP} \cdot P_{TPP}^{MAX} - P_t^{TPP}) > R_{TPP}^{UP} \end{cases};$$
$$\forall tpp \in TPP; \forall t \in T; \quad (2.27)$$

The maximum hourly power reserve capacity of TPPs for unloading:

$$R_t^{TPP\_DOWN} = \begin{cases} R_{TPP}^{SWITCH} \cdot \sum_{tpp}^{TPP} (P_t^{TPP} - Y_{ot}^{TPP} \cdot P_{TPP}^{MIN}) : R_{TPP}^{SWITCH} \cdot \sum_{tpp}^{TPP} (P_t^{TPP} - Y_{ot}^{TPP} \cdot P_{TPP}^{MIN}) \ge R_{TPP}^{DOWN} \\ R_{TPP}^{DOWN} : \qquad R_{TPP}^{SWITCH} \cdot \sum_{tpp}^{TPP} (P_t^{TPP} - Y_{ot}^{TPP} \cdot P_{TPP}^{MIN}) < R_{TPP}^{DOWN} \end{cases};$$
$$\forall tpp \in TPP; \forall t \in T. \quad (2.28)$$

Constraints on the use of import and export.
Constraints on the hourly use of import and export:

$$P_t^I \ge P_I^{MIN}; \quad P_t^I \le P_I^{MAX}; \forall t \in T; \quad (2.29)$$

$$P_t^E \ge P_E^{MIN}; \quad P_t^E \le P_E^{MAx}; \forall t \in T. \quad (2.30)$$

Constraints on the use of import and export over the course of a day:

$$\sum_t^T P_t^I \ge E_I^{D\_MIN}; \quad \sum_t^T P_t^I \le E_I^{D\_MAX}; \forall t \in T; \quad (2.31)$$

$$\sum_t^T P_t^E \ge E_E^{D\_MIN}; \quad \sum_t^T P_t^E \le E_E^{D\_MAX}; \forall t \in T. \quad (2.32)$$

Ensuring the balance of import and export volumes:

$$\begin{cases} \sum_t^T (P_t^I - P_t^E) \ge 0 : E_{IE}^D = 0 \\ \sum_t^T (P_t^I - P_t^E) = 0 : E_{IE}^D = 1 \end{cases}; \forall t \in T. \quad (2.33)$$

Restrictions on the operation of the energy storage system.

Constraints on the hourly charging and discharging capacity of the energy storage system:

$$E_t^{B\_C} = \sum_{1}^{t-1} \left(P_t^{B\_C} - P_t^{B\_G}\right) \le E_B^{D\_MAX}; \forall t \in T. \tag{2.34}$$

Restrictions on the operation of the energy storage system over the course of a day:

$$\sum_{t}^{T} P_t^{B\_C} \le E_B^{D\_MAX}; \quad \sum_{t}^{T} P_t^{B\_G} \le E_B^{D\_MAX}; \quad \forall t \in T. \tag{2.35}$$

Limitation on the power capacity of the energy storage system:

$$P_t^{B\_C} \le P_B^{C\_MAX}; \quad P_t^{B\_G} \le P_B^{G\_MAX}; \quad \forall t \in T. \tag{2.36}$$

Daily energy balance of the storage system—the sum of charging energy equals the sum of discharging energy:

$$\sum_{t}^{T} P_t^{B\_C} = \sum_{t}^{T} P_t^{B\_G}; \quad \forall t \in T. \tag{2.37}$$

Hourly power of the storage system's contribution to load balancing and unloading:

$$R_t^{B\_UP} = R_B^{SWITCH} \cdot \left(E_t^{B\_C} + P_t^{B\_C}\right); \forall t \in T; \tag{2.38}$$

$$R_t^{B\_DOWN} = R_B^{SWITCH} \cdot \left(P_t^{B\_G} + P_B^{C\_MAX}\right); \forall t \in T. \tag{2.39}$$

The operating limitations of HPP.

The minimum and maximum power values that the HPP supplies to the grid:

$$P_t^{WPP} \ge P_{WPP}^{MIN}; P_t^{WPP} \le P_{WPP}^{MAX}; \forall g \in WPP; \forall t \in T. \tag{2.40}$$

The minimum and maximum amount of energy that the HPP supplies to the grid throughout the day:

$$\sum_{t}^{T} P_t^{WPP} \ge E_{WPP}^{D\_MIN}; \quad \sum_{t}^{T} P_t^{WPP} \le E_{WPP}^{D\_MAX}; \quad \forall g \in WPP; \forall t \in T. \tag{2.41}$$

The maximum value of the HPP power change relative to the previous period:

$$P_t^{WPP} - P_{t-1}^{WPP} \leq P_{WPP}^{RUMP\_MAX}; P_{t-1}^{WPP} - P_t^{WPP} \leq P_{WPP}^{RUMP\_MAX}; \forall g \in WPP; \forall t \in T. \tag{2.42}$$

Ensuring overall reserve.

The limiting parameters of the overall reserve of the power system are technological constraints and are specified in the model according to [6].

Limitation of the hourly minimum reserve for the load:

$$R_t^{\Sigma\_UP} = R_t^{TPP\_UP} + R_t^{HP\_UP} + R_t^{B\_UP} \geq R_{ALL}^{MIN\_UP}; \forall t \in T. \tag{2.43}$$

Limiting the hourly minimum reserve for unloading:

$$R_t^{\Sigma\_DOWN} = R_t^{TPP\_DOWN} + R_t^{HP\_DOWN} + R_t^{B\_DOWN} \geq R_{ALL}^{MIN\_DOWN}; \forall t \in T. \tag{2.44}$$

The mathematical model [11] is implemented in the MathProg algebraic modeling language, which is part of the SolverStudio optimization package [18] and utilizes the COIN-OR PuLP modeling language. SolverStudio is an add-in for Excel 2007 and later versions for Windows, allowing the creation and solving of optimization models in Excel using several optimization modeling languages, including GMPL (GNU MathProg Language).

The main goal of developing mathematical model [11] is to create a tool for investigating the possibilities of optimized dispatching of generating capacities of TPPs and hydro units of pumped storage power plants (PSPPs), as well as generating and reserve capacities of large-scale energy storage systems (ESSs) within the Ukrainian power system. The model explores the permissible ranges of parameters for generating and storage capacities that allow for stable and balanced operation of the power system. Using the model enables determining optimal operating modes for generating and storage capacities to cover the schedule of electrical loads without utilizing storage, as well as with the use of storage under various total installed and peak capacities of SPP and WPP.

The multi-node integrated energy system model [10] with integer optimization variables for the composition and operating modes of generating, reserve, and storage capacities within a single synchronous zone, the continental Europe ENTSO-E, is also implemented in the GMPL algebraic modeling language on the SolverStudio platform. The model listing contains 1604 lines, while the model parameters file (Appendix B) contains 1124 lines.

The results of the calculated optimized energy balances of the multi-node integrated energy system for several typical load days are presented in tables and figures in the fourth chapter. Specifically, the results include calculations of generation by types of dispatchable technologies and volumes of cross-border flows.

2.5 Model for Minimizing the Weighted Average Cost of Production and Storage of Electricity

The model for minimizing the weighted average cost of electricity production using traditional methods, renewable energy sources (RES), and energy storage technologies [12] allows for assessing the potential reduction in the average hourly cost of electricity production. Based on forecasted daily load schedules, it solves the problem of predicting the optimal modes of operation for the generating and storage capacities of the system. Implementing the obtained solution ensures the minimization of the hourly cost of produced electricity in the IPS of Ukraine.

The model $\Omega(G, S, O, C, I, E, D, L)$ is characterized by the following set of key parameters, where for each generating technology j:

- $G_j\left(P^{\min}, P^{\max}, C, r\right)$—set of generating nodes, for each of which:
  - $P^{\min}, P^{\max}$—minimum and maximum generation capacity,
  - C_j—weighted average cost of electricity production,
  - r_j—set of technological constraints.
- $S_k\left(P^{\min}_{SCh}, P^{\max}_{SCh}, P^{\min}_{SDCh}, P^{\max}_{SDCh}, C_{Ch}, C_{DCh}, r\right)$—set of storage units, for each of which:
  - $P^{\min}_{SCh}, P^{\max}_{SCh}, P^{\min}_{SDCh}, P^{\max}_{SDCh}$—minimum and maximum power in charge (Ch) and discharge modes (DCh),
  - C_{Ch}, C_{DCh}—weighted average cost of utilization in charge and discharge modes,
  - r_k—technological constraints.
- I, E—similar parameters characterizing the import and export of electricity;
- D_T—load profile over the modeling horizon T.

The condition for covering the load schedule at each step t, $\forall t \in T$:

$$P_t^{NPP} + P_t^{CHP} + P_t^{PV} + P_t^{WIND} + P_t^{TPP} + P_t^{HPP} + P_t^{HPSS\_G} - P_t^{HPSS_P} + Z_t^I - Z_t^E = D_t. \tag{2.45}$$

So, in covering the load schedule, the following are involved:

- Total generation capacity of the following technologies:

$$P_t^{NPP} - \text{NPP}, P_t^{CHP} - \text{CHP}, P_t^{TPP} - \text{TPP}, P_t^{HPP}\text{HPP}, P_t^{PV} - \text{SPP},$$
$$P_t^{WIND}\text{WPP}, P_t^{HPSS\_G}\text{HPSS\_GEN}.$$

- Total power consumption of pumped storage HPPs—$P_t^{HPSS\_P}$;
- Total import Z_t^I and export Z_t^E capacity.

The total weighted average cost of *actual* electricity production in the IPS of Ukraine C^{Σ} is calculated by the following formula:

$$C^{\Sigma} = \sum_{t}^{T} \begin{pmatrix} C_{NPP}P_t^{NPP} + C_{TEC}P_t^{TEC} + C_{PV}P_t^{PV} + C_{WIND}P_t^{WIND} + C_{TPP}P_t^{TPP} + \\ C_{HPP}P_t^{HPP} + C_{HPSS\_G}P_t^{HPSS\_G} + C_{HPSS\_P}P_t^{HPSS\_P} + C_I Z_t^I - C_E Z_t^E \end{pmatrix}. \tag{2.46}$$

In the calculations of the *potential optimized* weighted average cost of electricity production in the IPS of Ukraine C^{Σ_O}, a target function has been used that ensures the minimum total weighted average cost:

$$C^{\Sigma\_O} = \sum_{t}^{T} \begin{pmatrix} C_{NPP}P_t^{NPP\_O} + C_{TEC}P_t^{TEC\_O} + C_{PV}P_t^{PV} + C_{WIND}P_t^{WIND} + \\ C_{TPP}P_t^{TPP\_O} + C_{HPP}P_t^{HPP\_O} + C_{HPSS\_G}P_t^{HPSS\_G\_O} + \\ C_{HPSS\_P}P_t^{HPSS\_P\_O} + C_I Z_t^{I\_O} - C_E Z_t^{E\_O} \end{pmatrix} \Rightarrow \min. \tag{2.47}$$

The model also takes into account technological constraints.

Model implemented using GLPK (GNU Linear Programming Kit), designed for solving linear, mixed-integer programming, and similar classes of mathematical problems [19], integrated into a software-information complex [20, 21]. Model calculations were performed using actual load schedules of the Unified Energy System of Ukraine in 2018 [21–25]. The calculation results, presented in the fourth chapter, confirmed the adequacy of the model and its usability for assessing the optimization potential of the generating resources of the IPS of Ukraine. Numeric estimates of the optimization potential of the operating modes of the components of the IPS of Ukraine were obtained using the model. In the optimized mode, the cost of produced electricity is reduced. By aligning the operation schedules of power units, the requirements for maneuvering modes are relaxed, and operational and technological characteristics are improved. The likelihood of the need for emergency repairs decreases, and the time between emergency and planned maintenance increases.

The model can serve as a basis for further development of a family of optimization models for the structure of generating and storage capacities, as well as for the further modernization and development of software-information systems for modeling the IPS of Ukraine.

2.6 Conclusions

1. Using a *new econometric-mathematical model of innovation diffusion in energy*, scenarios for the development of solar energy systems in Germany and Ukraine were calculated.

2. *The method of economic-technological forecasting of optimal operation and development parameters of the energy system, along with the model of long-term technological updating of the energy system structure*, allowed making forecasts for the consumption and generation structure of electricity in the IPS of Ukraine until 2040. These forecasts enable evaluating the potential for improving the components of the Ukrainian energy system.
3. *The model of hierarchically controlled quasi-dynamic energy systems* solves management tasks for complex dynamic energy systems. Specific cases of this model include the developed *model of a multi-node integrated energy system* and a *model for minimizing the weighted average cost of electricity production.*
4. *The multi-node integrated energy system model* solves optimization tasks for loading generating capacities of energy systems with the criterion of minimizing production costs and import/export of electricity.
5. *The model for minimizing the weighted average cost of electricity production* using traditional, renewable energy sources, and energy storage technologies allows assessing the potential reduction in the average hourly cost of electricity production. Based on forecasted daily load schedules, it solves the problem of predicting the optimal modes of operation for the generating and storage capacities of the system.

References

1. Kijek, A., Kijek, T.: Modelling of innovation diffusion. Oper. Res. Decis. **3–4**(20), 53–68 (2010)
2. Bass, F.M.: A new product growth for model consumer durables. Manag. Sci. **15**(5), 215–227 (1969). https://doi.org/10.1287/mnsc.15.5.215. Date of Access: 27 July 2023
3. Denysov, V., et al.: Improving the model of long-term technological update of power system components. Syst. Res. Energy. **2023**(2), 30–37 (2023). https://doi.org/10.15407/srenergy2023.02.030. Date of Access: 21 July 2023
4. Denisov V. Family of hierarchical models of sequential optimization of the energy systems sustainable development. In: Renewable Energy and Energy Efficiency in the 21st Century: Materials of the XVIII International Conference, pp. 119–123, Kyiv (2017)
5. Denysov, V.: Stochastic Quasi-dynamic model of economic-technological impact of the life cycle of innovative technologies. In: Problems of Regulatory Support for Innovative Activity and Ways to Solve Them, Kyiv, September 27, 2017. Kyiv, 2017. pp. 35–37 (2017)
6. Levelized cost of electricity and levelized avoided cost of electricity methodology supplement. Homepage—U.S. Energy Information Administration (EIA). https://www.eia.gov/renewable/workshop/gencosts/pdf/methodology_supplement.pdf. Date of Access: 21 July 2023
7. Levelized cost and levelized avoided cost of new generation resources in the annual energy outlook 2017. U.S. Energy Information Administration, 21 p (2017)
8. Marín-Cano, C., Mejía-Giraldo, D.: Levelized avoided cost of electricity model based on power system operation. DYNA **85**(206), 79–84 (2018). Home Page. https://doi.org/10.15446/dyna.v85n206.69577. Date of Access: 27 Oct 2023
9. Kostenko, G., Zaporozhets, A.: Transition from Electric Vehicles to Energy Storage: Review on Targeted Lithium-Ion Battery Diagnostics. **Energies**. 17(20), 5132 (2024). https://doi.org/10.3390/en17205132

10. Denisov, V.: Integrated power system multi-node model, taking into account the nondispatchable of renewable energy sources. In: 2022 IEEE 8th International Conference on Energy Smart Systems (ESS), Kyiv, Ukraine, 12–14 October 2022. 2022. https://doi.org/10.1109/ess57819.2022.9969255. Date of Access: 27 July 2023
11. Denisov, V.A.: Determination of optimal operating modes of the Ukrainian power system when covering the daily schedule of electrical loads, ensuring the necessary volumes of redundancy and using storage capacities. Probl. Gen. Energy **2020**(4), 33–44 (2020). https://doi.org/10.15407/pge2020.04.033. Date of Access: 27 July 2023
12. Denisov, V., Kulyk, M., Babak, V., Zaporozhets, A., Kostenko, G.: Modeling Nuclear-Centric Scenarios for Ukraine's Low-Carbon Energy Transition Using Diffusion and Regression Techniques. **Energies**. 17(20), 5229 (2024). https://doi.org/10.3390/en17205229
13. Regulation on Approval of the Transmission System Code. Decision of the National Commission for State Regulation of Energy and Public Utilities No. 309 of March 14, 2018. Updated as of April 14, 2023. Available at: https://zakon.rada.gov.ua/laws/show/v0309874-18#Text. Accessed: 27 July 2023
14. Shulzhenko, S.V., Turutikov O.I., Tarasenko P.V.: Model of mathematical programming with integer variables for determining the optimal regime of loading of hydroelectric pumped storage power plants for balancing daily profile of electric loads of the power system of Ukraine. Probl. Gen. Energy **2019**(4), 13–23 (2019). https://doi.org/10.15407/pge2019.04.013. Date of Access: 21 July 2023
15. Shulzhenko, S.V., Turutikov, O.I., Ivanenko N.P.: Mixed-integer linear programming mathematical model for founding the optimal dispatch plan of Ukrainian thermal power plants' units and hydro pumping storages stations' units for balancing daily load profile of power system of Ukraine. Probl. Gen. Energy **2020**(1), 14–23 (2020). https://doi.org/10.15407/pge2020.01.014. Date of Access: 21 July 2023
16. Sirota, I., et al.: Problems and prospects of operation of HPPs and PSPPs in the new electricity market. Hydroenergy of Ukraine **3**(4), 17–19 (2019)
17. Yevsyuk, M.M.: Electrical Stations and Substations: Lecture Notes for Students of Specialty 141 "Electrical Power Engineering, Electrical Engineering, and Electromechanics" (Full-Time Studies), 210 p. Technical College of Lutsk NTU, Lutsk (2018)
18. An integrated environment for optimisation using modelling languages within Excel. SolverStudio for Excel. https://solverstudio.org/download-install/. Date of Access: 29 July 2023
19. GLPK (GNU Linear Programming Kit): Linear Programming and Algorithms for Communication Networks, pp. 39–44 (2012). https://doi.org/10.1201/b12733-5. Date of Access: 01 Aug 2023
20. Denysov, V., Babak, V.: Software and information simulation complex of multi-node integrated and autonomous power and heat supply systems. Syst Res Energy. **2023**(3), 50–63. https://doi.org/10.15407/srenergy2023.03.050. Date of Access: 27 Sep 2023
21. Denysov, V.: Software and information complex for district heat supply systems modeling. Syst. Res. Energy **2022**(1), 38–45 (2022). https://doi.org/10.15407/srenergy2022.01.038. Date of Access: 27 Sep 2023
22. Hourly power balance of Ukraine's IPS for 2018. https://data.gov.ua/dataset
23. Hourly electricity balance of the IPS of Ukraine—Energy Map. Energy Map. https://map.ua-energy.org/en/resources/8998f2ed-379f-4fa9-9076-88782b32ee4f/. Date of Access: 21 July 2023
24. Daily Production/Consumption Schedule of Electricity. NPC Ukrenergo. https://ua.energy/diyalnist/dyspetcherska-informatsiya/dobovyj-grafikvyrobnytstva-spozhyvannya-e-e/. Accessed: 6 Oct 2020
25. Hourly electricity balance of the IPS of Ukraine—Energy Map. Energy Map. https://map.ua-energy.org/en/resources/8998f2ed-379f-4fa9-9076-88782b32ee4f/. Date of Access: 07 June 2023

Chapter 3
Software and Information Complex for Modeling Integrated Multi-node and Autonomous Electric and Heat Supply Systems

3.1 Composition and Structure of the Software and Information Complex

As already noted in Chap. 1, the development of theoretical foundations, methods, and tools for assessing the efficiency of functioning and development of energy systems, as well as mathematical methods and models for determining the capabilities and limitations of updating elements of both autonomous and integrated energy systems, has been the focus of many Ukrainian and foreign scientists. The consideration of the tasks and difficulties in solving them, as mentioned in these works, allows us to conclude the relevance and necessity of developing new and improving existing tools that combine the advantages of the mentioned approaches.

For practical research on improved and new approaches and their modeling, a new software and information complex (SIC) [1, 2] has been developed. The main advantages of the complex, which distinguish it from the known ones (Table 2 of the Chap. 1), include its versatility, allowing for the investigation of various energy system models in a short period of time. This versatility is ensured by the fact that the complex is developed using a combination of standard Microsoft Excel software and the SolverStudio add-in [3]—an add-in for Excel 2007 and later versions on Windows, which allows for the investigation of various optimization models using a wide range of optimization modeling languages:

- PuLP—open-source COIN-OR modeling language based on Python;
- COOPR/Pyomo, which extends Pulp with abstract models, support for stochastic programming and a wider range of solvers;
- AMPL—modeling language that supports running models in the cloud;
- GMPL (GNU MathProg Language)—open-source analogue of AMPL, developed as part of the GLPK (GNU Linear Programming Kit). GMPL is part of SolverStudio;
- GAMS—commercial modeling language;

A. Zaporozhets et al., *Structure Optimization of Power Systems with Renewable Energy Sources*, Studies in Systems, Decision and Control 583,
https://doi.org/10.1007/978-3-031-83697-8_3

- Gurobi—commercial solver available in SolverStudio;
- CMPL—an open-source COIN-OR modeling language.
- SimPy—open-source Python modeling language that is included in the SolverStudio download.

Thanks to the presence of the SolverStudio add-in [3] in the software and information complex [1, 2], the user can develop, edit, save and debug the optimization model in an Excel workbook. The connection of source data, sets, parameters, constants and variables used in the model is conveniently organized. It is possible to edit the parameters and source data, and then run the model. Simulation results can be displayed on model worksheets or output as separate files. Another advantage of the software and information complex [1, 2] is the ability to conveniently compare many models, due to the fact that each of the worksheets can have its own model.

The complex allows for the optimal selection of energy units and their operating modes, ensuring the production and redistribution of energy according to the consumption schedule.

The structure of the information and software complex is presented in Fig. 3.1.

The complex includes interacting blocks: service, control, information and analytical.

The *service block* is designed to improve the convenience of working with SIC and user service. It includes internal interface and user interface programs, control and dialog forms, provides opportunities for: generating statistical reports on specified

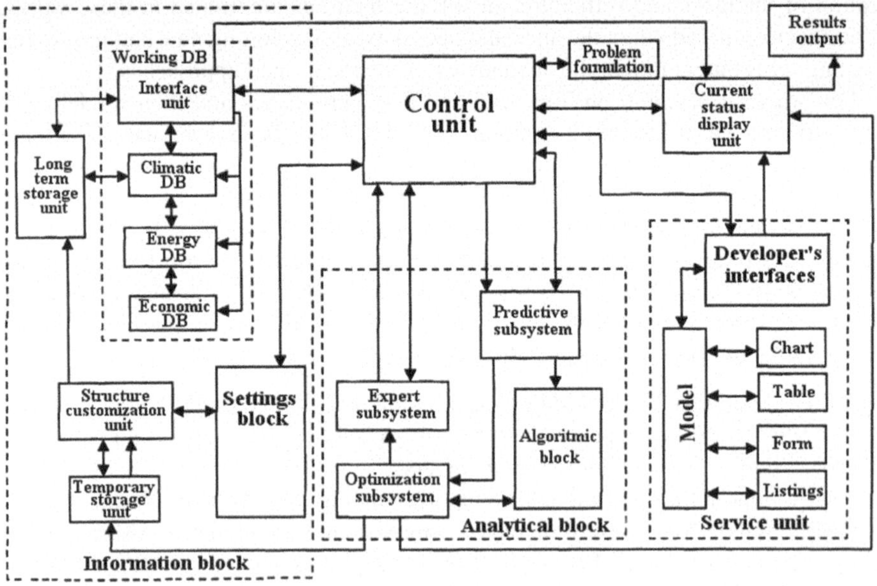

Fig. 3.1 Information and software complex's structure

features, visualization of intermediate data, parameters of the current state, results of calculations on request in a form convenient for the developer and user.

The main functions of the *control unit* are to coordinate the processes of entering initial data, optimizing and monitoring the results obtained.

The *information block* is designed for collecting, storing and primary processing of data and includes a set of databases (DB) that provide permanent storage and replenishment of data necessary for the functioning of the SIC, supports standard data management functions:

- input and editing;
- organization of storage;
- connection (download) and disconnection (unloading) of databases necessary for modeling at the current stage of work.

The information block, in addition to storing elements of the working database, allows you to perform the functions of selecting modes and variable parameters of the optimization process.

The analytical block is designed for solving tasks related to intelligent data processing and includes capabilities for preparing values for forecasting scenario calculations based on expert data [4–8], data from open sources [9–14], and others. It also allows for the selection of parameters for the optimization subsystem. Utilizing these capabilities enables the assessment of the impacts of various input factors on the hierarchical managed energy system based on the scenario method. The optimization subsystem allows for solving the tasks of selecting the best management options under given constraints.

3.2 Service Block

The example of the title page of the SIC with the dialog form of the service block is shown in Fig. 3.2. The dialog form of the service block (Fig. 3.3) allows for selecting or modifying the possible configuration of the modeling process.

The model text can be edited, and the optimization process can be observed using panel shown in Fig. 3.4.

The parameters of the model processing program, defined using GMPL (GNU MathProg Language) [15], can be edited using the form shown in Fig. 3.5 if they need modification. Alternatively, these parameters can be left unchanged if the user wants to keep the default values. For example, increasing the value of the "mipgap" parameter, which affects the accuracy of finding an integer solution, can reduce the search time.

Parameters that are most frequently subject to change, such as the modeling date, scaling coefficients, and others, are conveniently placed in a separate parameters form (Fig. 3.6).

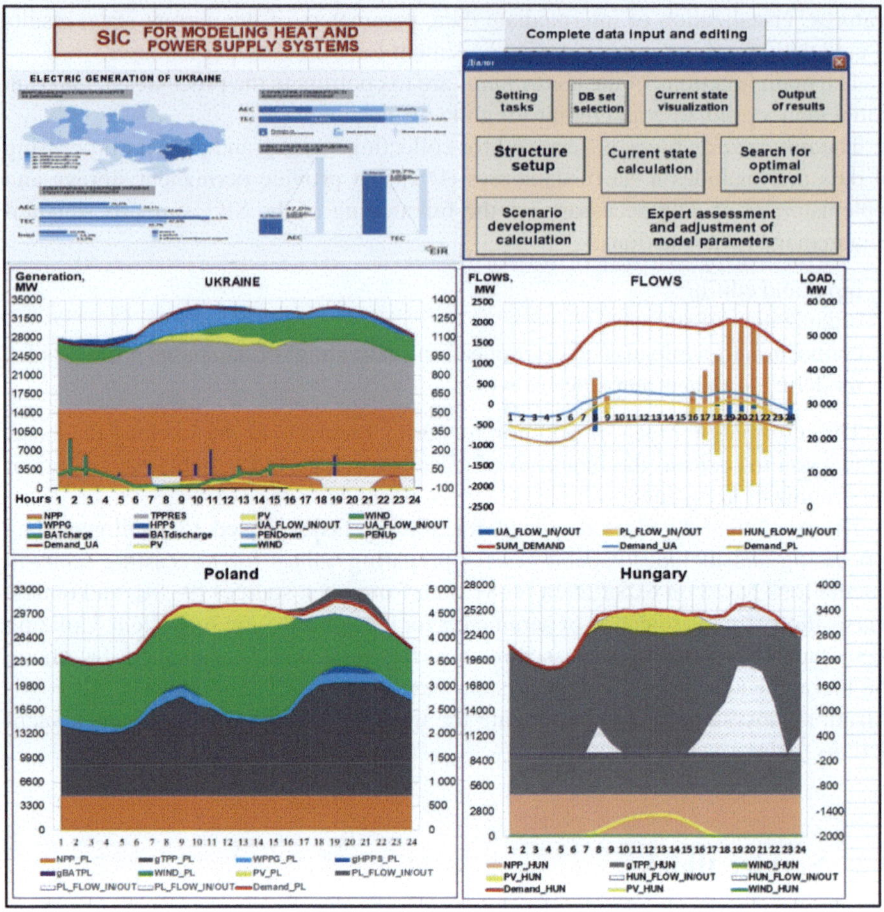

Fig. 3.2 Title page of the SIC with the dialog form of the service block

Fig. 3.3 Dialog form of the service block

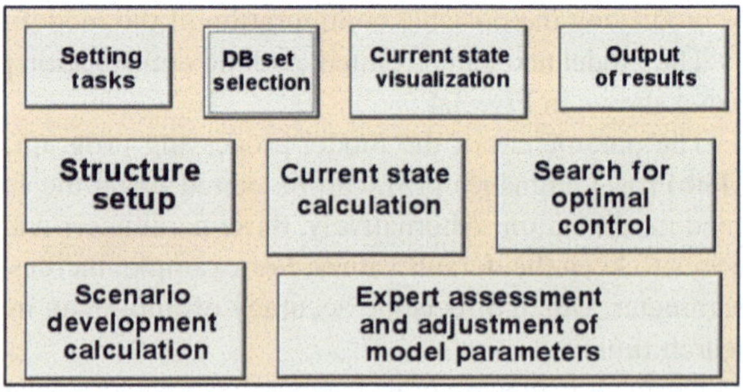

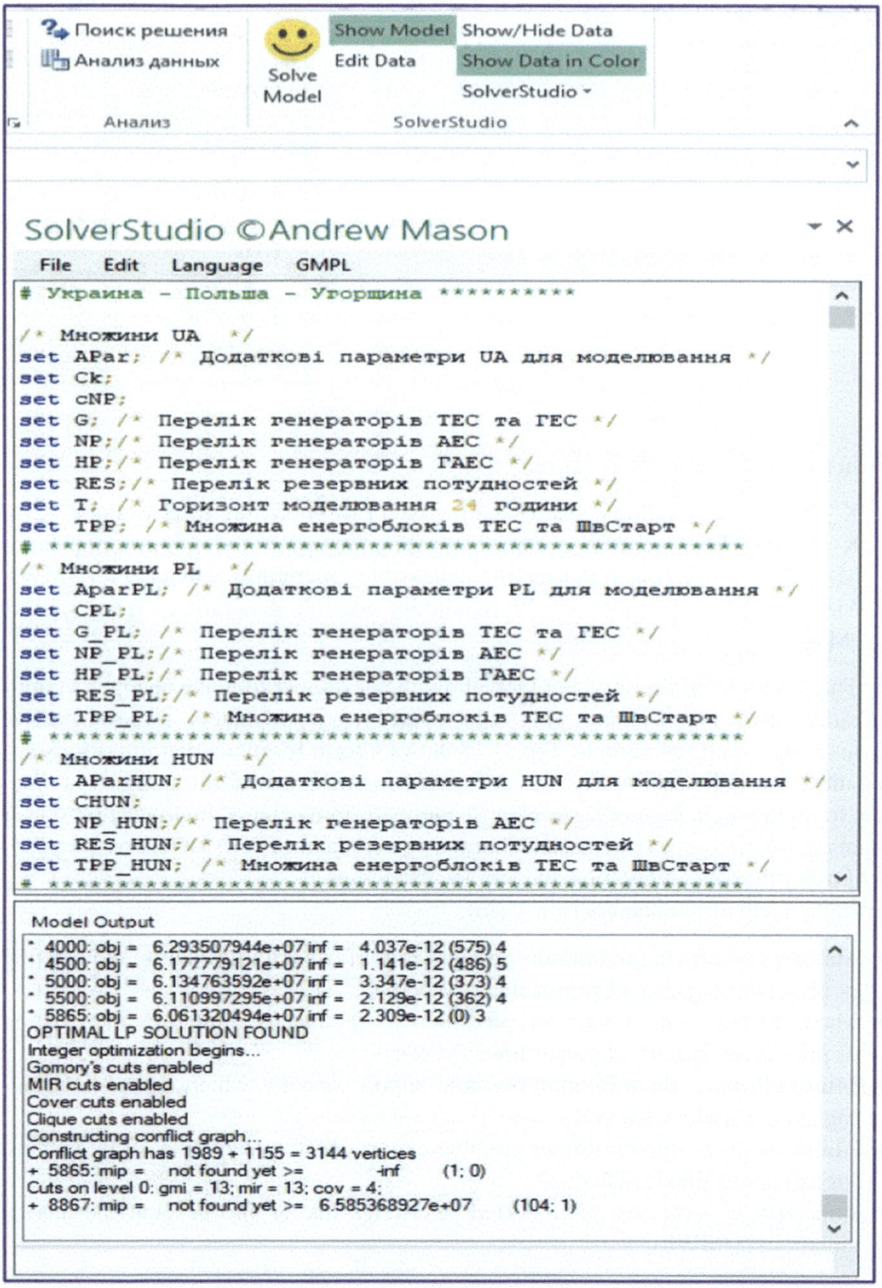

Fig. 3.4 The panel for model text editing and observation of the optimization process

GMPLOptions	
cover	
clique	
gomory	
mir	
mipgap	0.05
m	

Fig. 3.5 Form for selecting parameters of the model processing program

In the form of parameters (Fig. 3.6):

- K_PV—PV generating capacity scaling factor for modeling forecast data;
- K_WIND—WPP generating capacity scaling factor for modeling forecast data;
- MAX PV—limitation of the maximum possible generating capacity of PV, (MW);
- MAX WIND—limitation of the maximum possible generating capacity WPP, (MW).

The functions of the service block allow the developer to make changes to optimization algorithms by modifying values and binary parameters in the respective forms (Fig. 3.7). For example, Fig. 3.7a shows a form for managing import–export parameters. Additionally, Fig. 3.7 presents forms for modifying control parameters for the participation of hydroelectric pumped storage units, hydroelectric power plants (HPPs), storage systems, and thermal power plants (TPPs) for reserve purposes [16] using the total generating and storage capacities.

In the form of parameters (Fig. 3.7a):

- MaxImpPower—the maximum possible import capacity limited by the existing or projected capacity of power lines (MW);
- MaxExpPower—the maximum possible export capacity limited by the existing or projected capacity of power lines (MW);
- MinImpPower—the minimum possible import capacity, which is set by technological constraints, (MW);
- MinExpPower—the minimum possible export capacity, which is set by technological constraints, (MW);
- ImportPrice—unit cost of imported electricity, taking into account the market situation, ($/MWh);

DATE	K_PV	K_WIND	MAX PV	MAX WIND
February 28, 2040	16.393	19.444	11000	7000

Fig. 3.6 Parameters editing form

MaxImpPower	650
MaxExpPower	650
MinImpPower	0
MinExpPower	0
ImportPrice	200
ExportPrice	200
DailyBalanceExpImp	0

a

EachHydroPump	1
PumpGenBalance	1
MaxSumPumpPower	1100
MaxSumGenPower	1100
PumpGenOperate	1
On_Off_HPPS_Reseve	1
Include_HPPS_UpReseve	1
Include_HPPS_DownReseve	1
Up_HPPS_ReserveLevel	3500
Down_HPPS_ReserveLevel	180

b

On_Off_WPP_Reseve	1
Include_WPP_UpReseve	1
Include_WPP_DownReseve	1
Up_WPP_ReserveLevel	500
Down_WPP_ReserveLevel	250

c

On_Off_BUT_Reseve	1
BatteryBalance	1
BatChargePrice	37
BatDisChargePrice	37
DailyMaxStorage	3000
Max_CHARGE_SumStoragePower	680
Max_DISCHARGE_SumStoragePower	3060

d

On_Off_TPP_Reseve	1
Include_TPP_UpReseve	1
Include_TPP_DownReseve	1
RampUp_TPP_ReserveLevel	300
RampDown_TPP_ReserveLevel	200

e

On_Off_ALL_Reseve	1
Include_ALL_UpReseve	0
Include_ALL_DownReseve	0
RampUp_ALL_ReserveLevel	1000
RampDown_ALL_ReserveLevel	500

f

Fig. 3.7 Forms for managing import–export parameters, as well as parameters for controlling the participation of hydroelectric pumped storage units, HPPs, storage systems, and TPPs in reserve operations: **a**—import–export; **b**—hydroelectric pumped storage units; **c**—HPPs; **d**—storage systems; **e**—TPPs; **f**—total reserve operations

- ExportPrice—unit cost of exported electricity, taking into account the market situation, ($/MWh);
- DailyBalanceExpImp—binary variable that determines whether the simulation will fulfill the condition: 1—zero balance of flows, 0—do not comply with the condition of zero balance of flows.

In the form of parameters (Fig. 3.7b):

- EachHydroPump—binary variable that determines whether the simulation will fulfill the condition: 1—each hydraulic unit of the PSP must be started once, 0—the condition is not met;
- PumpGenBalance—binary variable that determines whether the simulation will fulfill the condition: 1—each hydraulic unit of the PSP must use all the accumulated water for the current day, 0—the condition can be ignored;
- MaxSumPumpPower—the maximum allowable level of power consumption for injection by all PSPs, (MW);

- MaxSumGenPower—the maximum allowable level of power generation by all PSPs, (MW);
- PumpGenOperate—binary variable that determines whether the simulation will fulfill the condition: 1—cannot be simultaneously pumped and generated, if in a certain period of time the hydraulic unit pumps then all niches of the hydraulic aggregate cannot generate, 0—it is possible;
- On_Off_HPPS_Reserve—binary variable that determines whether the simulation will fulfill the condition: 1—use PSP capacities in common redundancy, 0—do not use;
- Include_HPPS_UpReserve—binary variable that determines whether the simulation will fulfill the condition: 1—use PSP capacities in reserve for load, 0—do not use;
- Include_HPPS_DownReserve—binary variable that determines whether the simulation will fulfill the condition: 1—use PSP capacities in redundancy for unloading, 0—do not use;
- Up_HPPS_ReserveLevel—the maximum possible capacity of participation of PSPs in reserve for load, (MW);
- Down_HPPS_ReserveLevel—the maximum possible capacity of participation of PSPs in reservation for unloading, (MW).

In the form of parameters (Fig. 3.7c):

- On_Off_WPP_Reserve—binary variable that determines whether the simulation will fulfill the condition: 1—use HPP capacities in common redundancy, 0—do not use;
- Include_WPP_UpReserve—binary variable that determines whether the simulation will fulfill the condition: 1—use HPP capacity in reserve for load, 0—do not use;
- Include_WPP_DownReserve—binary variable that determines whether the simulation will fulfill the condition: 1—use HPP capacity in redundancy for unloading, 0—do not use;
- Up_WPP_ReserveLevel—the maximum possible capacity of participation of HPPs in the reservation for loading, (MW);
- Down_WPP_ReserveLevel—the maximum possible capacity of participation of HPPs in reservation for unloading, (MW).

In the form of parameters (Fig. 3.7d):

- On_Off_BUT_Reserve—binary variable that determines whether the simulation will fulfill the condition: 1—use the capacity of the drives in the general redundancy, 0—do not use;
- BatteryBalance—binary variable that determines whether the simulation will fulfill the condition: 1—battery charge must be greater than the discharge in the previous hours, 0—battery charge must be greater than zero;
- BatChargePrice—unit cost of battery charge, ($/MWh);
- BatDisChargePrice—unit cost of battery discharge, ($/MWh);

- DailyMaxStorage—the maximum possible amount of charge–discharge of drives, (MWh per day);
- Max_CHARGE_SumStoragePower—the maximum possible total power consumed when charging the drives (MW);
- Max_DISCHARGE_SumStoragePower—the maximum possible total power given by all drives to the system (MW).

In the form of parameters (Fig. 3.7e):

- On_Off_TPP_Reserve—binary variable that determines whether the simulation will fulfill the condition: 1—use TPP capacities in general redundancy, 0—do not use;
- Include_TPP_UpReserve—binary variable that determines whether the simulation will fulfill the condition: 1—use TPP capacities when reserving for load, 0—do not use;
- Include_TPP_DownReserve—binary variable that determines whether the simulation will fulfill the condition: 1—use the capacity of TPPs when reserving for unloading, 0—do not use;
- RampUp_TPP_ReserveLevel—minimum permissible capacity of TPP participation in load redundancy, (MW);
- RampDown_TPP_ReserveLevel—minimum permissible capacity of TPP participation in reservation for unloading, (MW).

In the form of parameters (Fig. 3.7f):

- On_Off_ALL_Reserve—binary variable that determines whether the simulation will fulfill the condition: 1—use the power system capacity in redundancy, 0—do not use;
- Include_ALL_UpReserve—binary variable that determines whether the simulation will fulfill the condition: 1—use the power of the system when redundant for the load, 0—do not use;
- Include_ALL_DownReserve—binary variable that determines whether the simulation will fulfill the condition: 1—use the capacity of the system when redundant for unloading, 0—do not use;
- RampUp_ALL_ReserveLevel—minimum permissible capacity of general reserve for load, (MW);
- RampDown_ALL_ReserveLevel—minimum permissible capacity of the total reservation for unloading, (MW).

The methods of displaying output parameters and calculation results can be varied by editing the model text (Fig. 3.4). The convenience of visualizing input data and obtained results is ensured by the respective visualization and result output software blocks.

3.3 Information Block

As noted above, the information block is designed for collecting, storing, and initial processing of data and includes a set of databases (DBs) that ensure the constant storage and replenishment of the data necessary for the functioning of the SIC. It supports standard data management functions: data entry and editing, storage organization, connecting (loading) and disconnecting (unloading) the databases required for modeling at the current stage of work. Depending on the task being solved and the type of energy system being modeled (electric, thermal, autonomous, integrated), the necessary databases from the set of databases are connected or disconnected (Fig. 3.8).

In Fig. 3.8 in the columns are displayed:

- TIME_ALL—hour number since the beginning of the year;
- DATE_ALL—date and hour.

The following columns reflect the actual hourly values [17–20]:

- DEMAND_ALL_PL—actual hourly load schedule of the EC in Poland;
- DEMAND_ALL_HUN—actual hourly load schedule of the Hungarian EC;
- DEMAND_ALL_ROM—actual hourly load schedule of the EC in Romania;
- PV_PL—actual hourly schedule of capacity of PV in Poland;
- WIND_PL—actual hourly capacity schedule of wind farms in Poland;
- PV_HUN—actual hourly capacity schedule of Romanian PV.

TIME_ALL	DATE_ALL	DEMAND_ALL_PL	DEMAND_ALL_HUN	DEMAND_ALL_ROM	PV_PL	WIND_PL	PV_HUN
1392	27.02.2018 23:00	21106	5589	7779	0	0	0
1393	28.02.2018 0:00	20025	5333	7457	0	1935	0
1394	28.02.2018 1:00	19365	5049	7271	0	1888	0
1395	28.02.2018 2:00	19063	4840	7183	0	1879	0
1396	28.02.2018 3:00	19106	4731	7164	0	1987	0
1397	28.02.2018 4:00	19474	4762	7273	0	2150	0
1398	28.02.2018 5:00	20350	5058	7457	0	2168	0
1399	28.02.2018 6:00	22455	5675	7969	11	2178	13
1400	28.02.2018 7:00	24431	6184	8661	176	2205	75
1401	28.02.2018 8:00	25582	6252	9273	507	2138	169
1402	28.02.2018 9:00	25806	6306	9511	845	2147	252
1403	28.02.2018 10:00	25604	6297	9509	1061	2243	298
1404	28.02.2018 11:00	25759	6350	9366	1063	2611	327
1405	28.02.2018 12:00	25766	6322	9453	979	2752	335
1406	28.02.2018 13:00	25700	6296	9297	913	2882	313
1407	28.02.2018 14:00	25387	6191	9011	722	2893	243
1408	28.02.2018 15:00	24903	6227	8842	576	2739	143
1409	28.02.2018 16:00	24737	6161	8820	188	2470	37
1410	28.02.2018 17:00	25435	6192	8894	18	2383	2
1411	28.02.2018 18:00	26258	6497	9385	0	2452	0
1412	28.02.2018 19:00	26081	6537	9606	0	2497	0
1413	28.02.2018 20:00	25439	6374	9449	0	2348	0
1414	28.02.2018 21:00	23893	6183	9037	0	1946	0
1415	28.02.2018 22:00	22307	5875	8389	0	1604	0
1416	28.02.2018 23:00	20934	5686	7823	0	1404	0

Fig. 3.8 Element of the working DB, actual capacities (MW) of Poland, Hungary, Romania power systems

The information block, in addition to storing elements of the working database, contains matrices of economic, energy, and climatic parameters of the modeled energy systems (Figs. 3.9, 3.10 and 3.11). Changing the values of elements in the matrices of economic, energy, and climatic parameters allows for performing functions of selecting modes and parameters of the optimization process.

In the columns of the economic and energy parameters matrix of TPPs of Ukraine Fig. 3.9:

- TPP—list of types of power units of TPPs of Ukraine;
- pTPP Units—number of power units of each type;
- pTPP InstCap—installed capacity of the unit, (MW);
- pTPP UnitMax—maximum possible power that the power unit gives to the system (MW);

TPP	pTPP Units	pTPP InstCap	pTPP UnitMax	pTPP UnitMin	pTPP Unit Ramp	pTPP Unit Ramp Max	pTPP ProdCost	pTPP NumStart	pTPP MinUnits	pTPP MaxUnits	pTPP Time Operate
TPP800	1	800	750	500	20	100	55	1	1	1	24
TPP100	2	100	96	76	10	20	65	1	0	1	24
TPP150	2	150	139.5	96	10	40	63	1	0	1	24
TPP200	24	200	190	110	10	50	60	2	0	1	24
TPP300	22	300	280	175	10	100	58	2	5	12	24
FBCoal	0	1500	290	140	20	100	80	1	0	0	24
QSGas	0	0	9.43	0.1	9	9	95	1	0	0	24

Fig. 3.9 Matrix of economic and energy parameters of TPPs in Ukraine

NP	pNPP InstCap	pNPPT	pNPP ProdCost	pNPP MinUnits	pNPP MaxUnits
NPP1000	1000	24	22	3	13
NPP440	440	24	24	0	2

Fig. 3.10 Matrix of economic and energy parameters of NPPs of Ukraine

HP_number	HP	pP	pG	pHPGV	pHPPV	pHPT	pHPS	pHPVol
1	KHPS1	43	37	28.5	19	3	1	111
2	KHPS2	43	37	28.5	19	3	1	111
3	KHPS3	43	37	28.5	19	3	1	111
4	DHPS1	421	324	22.5	15	3	1	972
5	DHPS2	421	324	22.5	15	3	1	972
6	DHPS3	421	324	22.5	15	3	1	972
7	THPS1	216.5	151	25.5	17	3	1	453
8	THPS2	216.5	151	25.5	17	3	1	453

Fig. 3.11 Matrix of economic and energy parameters of pumped storage power plants in Ukraine

- pTPP UnitMin—minimum possible power that the power unit gives to the system (MW);
- pTPP Unit Ramp—accepted for calculations permissible change in the capacity of a TPP power unit when it is operating (MW);
- pTPP Unit Ramp Max—the maximum possible change in the capacity of a TPP power unit when it is operating (MW);
- pTPP ProdCost—the cost of supplying electricity to the grid by a TPP unit, ($/MWh);
- pTPP NumStart—the number of starts of TPP power units during the day;
- pTPP MinUnits—minimum number of TPP power units involved in covering the load schedule;
- pTPP MaxUnits—the maximum number of TPP power units that can participate in covering the load schedule;
- pTPP Time Operate—the permissible number of hours of operation of a TPP power unit during a day.

In the columns of the economic and energy parameters matrix of NPPs of Ukraine Fig. 3.10:

- NP—list of types of power units of Ukrainian NPPs;
- pNPP InstCap—installed capacity of the unit, (MW);
- pNPPT—number of hours of continuous operation of NPP units;
- pNPP ProdCost—the cost of supplying electricity to the grid by a nuclear power unit, ($/MWh);
- pNPP NumStart—number of NPP power units starts during the day;
- pNPP MinUnits—minimum number of NPP power units involved in covering the load schedule;
- pNPP MaxUnits—the maximum number of NPP power units allowed to participate in covering the load schedule.

In the columns of the matrix of economic and energy parameters of pumped storage power plants in Ukraine (Fig. 3.11):

- HP_number—sequence number of the hydro unit of the pumped storage power plant in Ukraine;
- HP—list of types of hydro units of the pumped storage power plants in Ukraine;
- pP—power consumed by the hydro unit of the pumped storage power plant in pumping mode (MW);
- pG—power of the hydro unit of the pumped storage power plant in generation mode (MW);
- pHPGV—costs of operating the hydro generator of the pumped storage power plant in generation mode;
- pHPPV—costs of operating the hydro generator of the pumped storage power plant in pumping mode;
- pHPT—number of hours of continuous operation of the hydro units of the pumped storage power plant in pumping and generation modes;

- pHPS—allowable number of starts of the hydro units of the pumped storage power plant in pumping and generation modes;

pHPVol—maximum amount of energy that can be stored by the hydro unit of the pumped storage power plant in pumping mode (MWh).

3.4 Analytical Block

Through the combined use and interaction of the service block, control block, and information block, the user prepares various information for the analytical block. The analytical block includes the built-in Solver Studio [3] optimizer. The optimization subsystem allows solving tasks for selecting the best management options for the energy system under given constraints. Once all the initial information is prepared, the user clicks the "Complete data input and editing" button (Fig. 3.2), and the control block creates a working data file and starts the optimization process. If the data is prepared correctly, the optimizer first finds an analog solution and then, considering the integrity and binarity of individual variables, an optimal solution. The calculation results are output by the visualization and result output blocks to the worksheets (Figs. 3.13, 3.14, 3.15, 3.16, 3.17 and 3.18) of the software and information complex. Otherwise, the optimizer issues a message indicating that no solution exists (Fig. 3.12).

The composition and structure of the use of mathematical models embedded in the software and information complex are presented in Fig. 3.19.

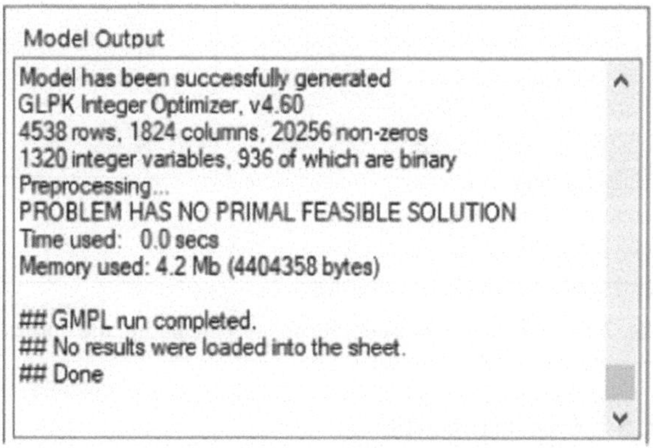

Fig. 3.12 Optimizer message indicating that no solution exists

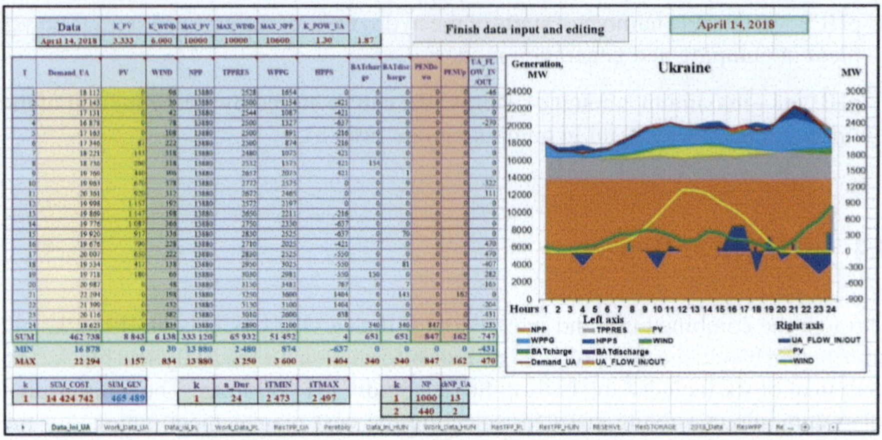

Fig. 3.13 Worksheet with the results of calculating the optimal parameters of the IPS of Ukraine

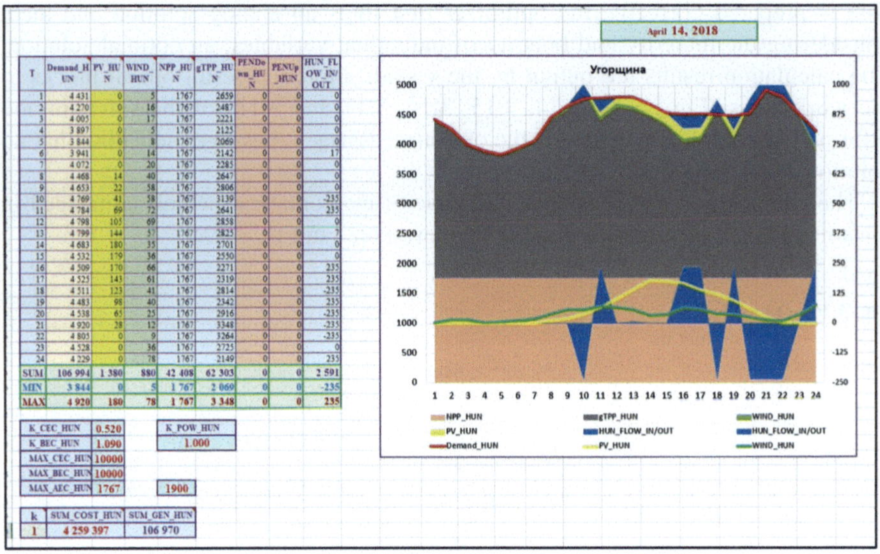

Fig. 3.14 Worksheet with the results of calculating the optimal parameters of the IPS of Poland

3.5 Conclusions

The developed SIC [1, 2] allows for detailed calculations of energy, technological, and economic indicators for the optimal use of energy system components. It determines the permissible limits of the operating parameters of the energy units in autonomous and integrated energy systems. Based on the results of these calculations, appropriate

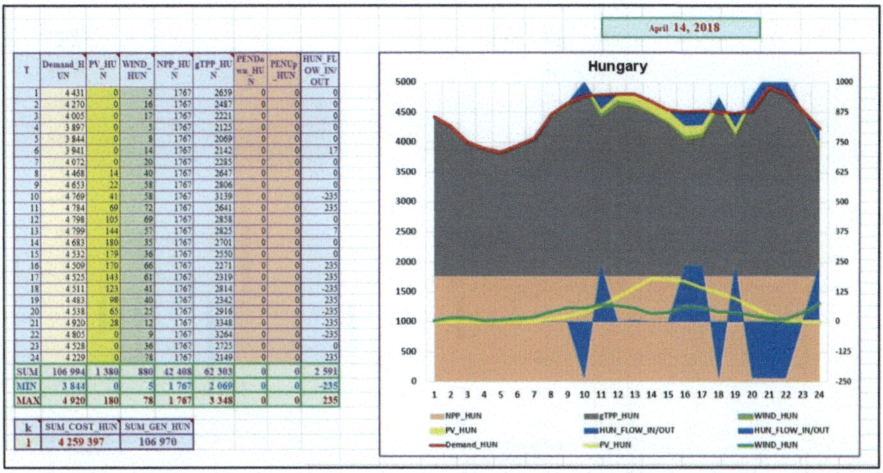

T	Demand_HUN	PV_HUN	WIND_HUN	NPP_HUN	gTPP_HUN	PENDown_HUN	PENUp_HUN	HUN_FLOW_IN/OUT
1	4 431	0	5	1767	2659	0	0	0
2	4 270	0	16	1767	2487	0	0	0
3	4 005	0	17	1767	2221	0	0	0
4	3 897	0	5	1767	2125	0	0	0
5	3 844	0	8	1767	2069	0	0	0
6	3 941	0	14	1767	2142	0	0	17
7	4 072	0	20	1767	2285	0	0	0
8	4 468	14	40	1767	2647	0	0	0
9	4 653	22	58	1767	2806	0	0	0
10	4 769	41	58	1767	3139	0	0	-235
11	4 784	69	72	1767	2641	0	0	235
12	4 798	105	69	1767	2858	0	0	0
13	4 799	144	57	1767	2825	0	0	7
14	4 683	180	35	1767	2701	0	0	0
15	4 532	179	36	1767	2550	0	0	0
16	4 509	170	66	1767	2271	0	0	235
17	4 525	143	61	1767	2319	0	0	235
18	4 511	123	41	1767	2814	0	0	-235
19	4 483	98	40	1767	2342	0	0	235
20	4 538	65	25	1767	2916	0	0	-235
21	4 920	28	12	1767	3348	0	0	-235
22	4 805	0	9	1767	3264	0	0	-235
23	4 528	0	36	1767	2725	0	0	0
24	4 229	0	78	1767	2149	0	0	235
SUM	106 994	1 380	880	42 408	62 303	0	0	2 591
MIN	3 844	0	5	1 767	2 069	0	0	-235
MAX	4 920	180	78	1 767	3 348	0	0	235

k	SUM_COST_HUN	SUM_GEN_HUN
1	4 259 397	106 970

Fig. 3.15 Worksheet with the results of calculating the optimal parameters of the IPS of Hungary

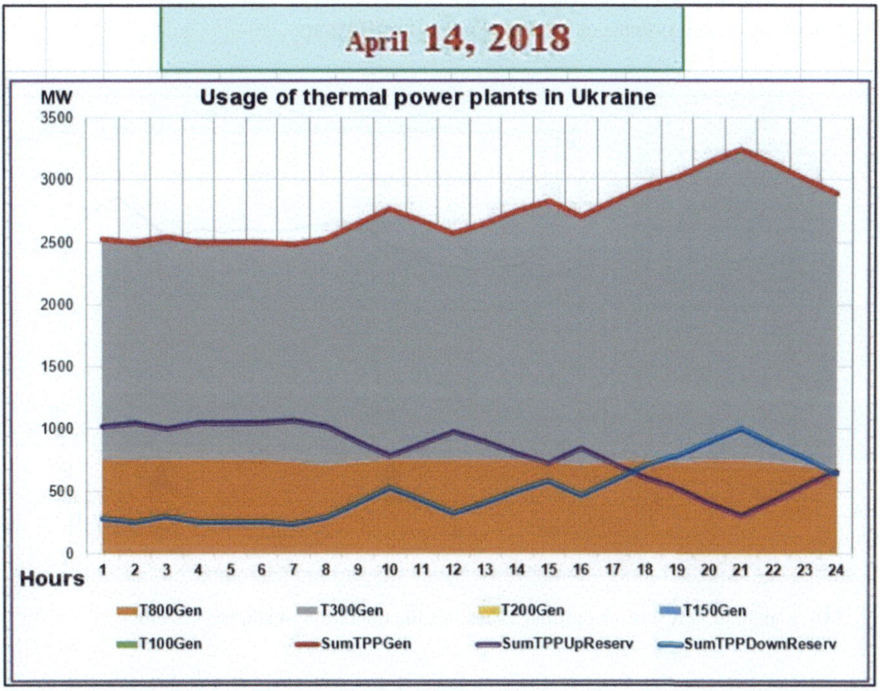

Fig. 3.16 Results of the calculation of the optimal use of TPP of Ukraine

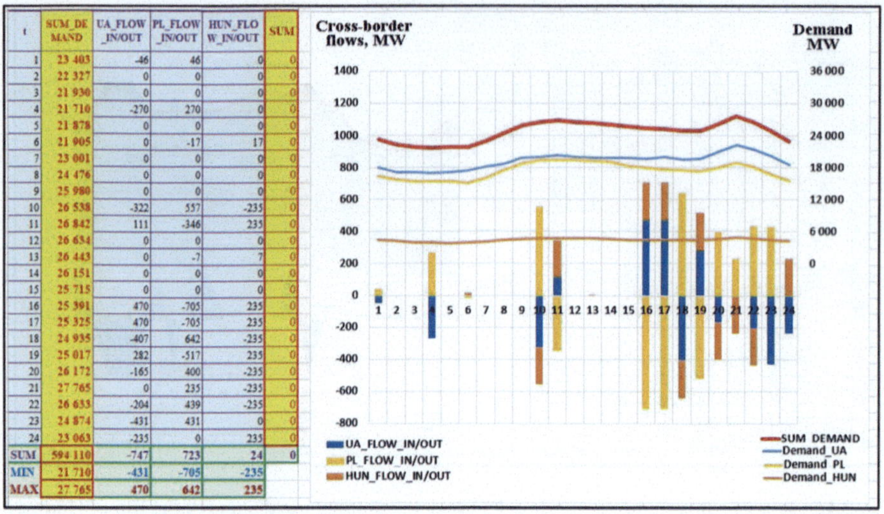

Fig. 3.17 Calculation results of optimal values of transboundary flows during the synchronous operation of the energy systems of Ukraine, Poland, and Hungary

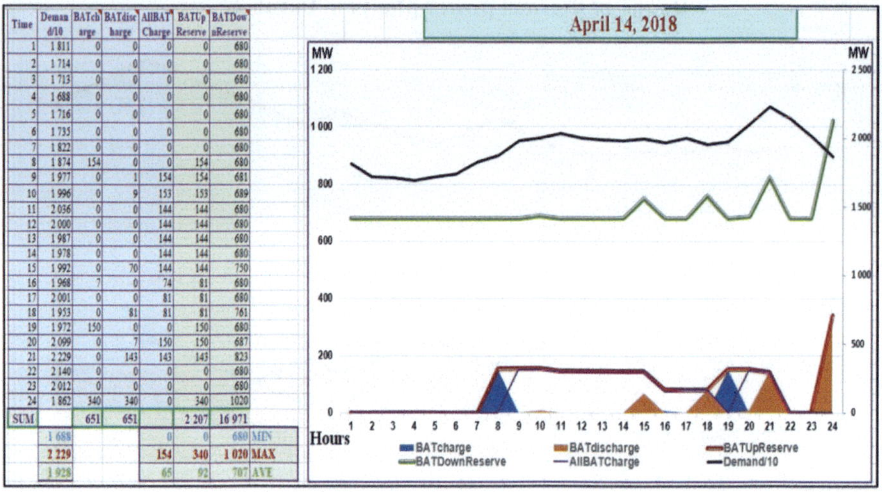

Fig. 3.18 Calculation results of optimal values for the operation of storage systems

measures for the prospective renewal of electricity and heat production technologies are selected.

The main advantages of the complex, which distinguish it from known ones (Table 2 of the Chap. 1), include its versatility, allowing for the study of various energy system models in a short time. This versatility is ensured by the fact that the complex is developed using a combination of standard software, Microsoft Excel,

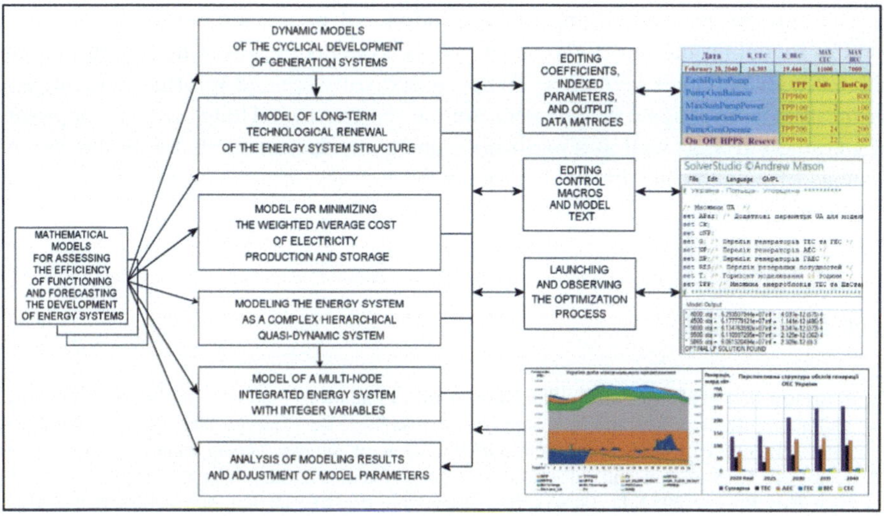

Fig. 3.19 The composition and structure of the use of mathematical models embedded in the software and information complex

and SolverStudio [3]—an add-in for Excel 2007 and later versions on Windows, which allows for the exploration of various optimization models using a wide range of optimization modeling languages.

Thanks to the SolverStudio [3] add-in, the information complex allows the user to develop, edit, save, and debug the optimization model directly in an Excel workbook. The connection of input data, sets, parameters, constants, and variables used in the model is conveniently organized. After editing the parameters and input data, the optimization calculation is launched. The modeling results can be displayed both on the model sheets and as separate files. Another advantage of the software and information complex is the ability to conveniently compare multiple models, as each worksheet can contain a different model.

The ability to include software implementations of various models of energy system operation in the software and information complex allows for the calculation of comparative characteristics of the most efficient technical, financial, and environmental parameters of technologies for all permissible operating modes in energy systems. This also enables the comparison of the efficiency of technologies and the need for their renewal, which is important for developing scenarios for the evolution of existing and new energy systems.

The main distinction of the developed software and information complex is the ability to thoroughly examine the impact of economic and technological parameters, as well as individual energy units and nodes within the energy system. These parameters can be presented either as matrices on the sheets of the software and information complex or as separate attached files available for automated input by the software and information complex.

The ease of use and editing of both individual parameters and software texts used in modeling the development of energy systems improves the quality of the obtained development scenarios. The proposed software and information complex can be used to investigate the prospects for the short-term and long-term development of Ukraine's IPS, as well as a multi-node energy system integrated with the energy systems of neighboring ENTSO-E member countries, which is relevant in current conditions.

References

1. Denysov, V., Babak, V.: Software and information simulation complex of multi-node integrated and autonomous power and heat supply systems. Syst. Res. Energy **2023**(3), 50–63 (2023). https://doi.org/10.15407/srenergy2023.03.050. Date of Access: 27 Sep 2023
2. Denysov, V.: Software and information complex for district heat supply systems modeling. Syst. Res. Energy. **2022**(1), 38–45 (2022). https://doi.org/10.15407/srenergy2022.01.038. Date of Access: 27 Sep 2023
3. An integrated environment for optimisation using modelling languages within Excel. Solver-Studio for Excel. https://solverstudio.org/download-install/. Date of Access: 29 July 2023
4. Lai, C.S., et al.: A review on long-term electrical power system modeling with energy storage. J. Clean. Prod. **280**, 124298 (2021). https://doi.org/10.1016/j.jclepro.2020.124298. Date of Access: 21 July 2023
5. Gaur, A.S., et al.: Long-term energy system planning considering short-term operational constraints. Energy Strat. Rev. **26**, 100383 (2019). https://doi.org/10.1016/j.esr.2019.100383. Date of Access: 21 July 2023
6. Denysov, V., Kulyk, M., Babak, V., Zaporozhets, A., Kostenko, G.: Modeling Nuclear-Centric Scenarios for Ukraine's Low-Carbon Energy Transition Using Diffusion and Regression Techniques. **Energies**. 17(20), 5229 (2024). https://doi.org/10.3390/en17205229
7. Report on the Assessment of the Adequacy of Generating Capacities to Meet Forecasted Electricity Demand and Ensure Necessary Reserves for 2022, approved by the NKREKP Resolution No. 664 dated April 11, 2023. NPC "Ukrenergo". Available at: https://ua.energy/zvit-z-otsinky-vidpovidnosti-dostatnosti-generuyuchyh-potuzhnostej/. Accessed: 21 July 2023
8. Maistrenko N.Y.: The prediction of energy consumption for the sector of general government in the Ukrainian economy until 2040. Probl. Gen. Energy. **2022**(1–2), 82–89 (2022). https://doi.org/10.15407/pge2022.01-02.082. Date of Access: 02 Aug 2023
9. Diatlova, V., Petryk, I.: Modern approaches to the development of renewable energy as a component of Ukraine's innovative economy. Econ. Organ. Manag. **2019**(2), 20–27 (2019). Available at: https://doi.org/10.31558/2307-2318.2019.2.2. Accessed: 1 Aug 2023
10. Kudria, S., et al.: Concept of the Roadmap for the Development of Hydrogen Energy in Ukraine until 2035. Renew. Energy **2019**(4), 22–28 (2019). Available at: https://doi.org/10.36296/1819-8058.2019.4(59).22-28. Accessed: 1 Aug 2023
11. Bila, S.O.: Development of Renewable Energy as a Component of Transition to Sustainable Development: Experience of the EU and Ukraine [Thesis] (2018). Available at: http://er.nau.edu.ua/handle/NAU/40118. Accessed: 1 Aug 2023
12. Savytskyi, O., & Aliieva, O. (2023). "Energy Transition"—The Biggest Challenge of Our Time for Ukraine. Ukrainian Climate Network | Working Group of Non-Governmental Environmental Organizations on Climate Change. Available at: https://ucn.org.ua/?p=4528. Accessed: 21 July 2023

13. Cabinet of Ministers of Ukraine.: Approval of the Energy Strategy of Ukraine for the Period up to 2035 Security, Energy Efficiency, Competitiveness: Order No. 605-p, August 18, 2017 (2017). Available at: https://zakon.rada.gov.ua/laws/show/605-2017-p#Text. Accessed: 7 Aug 2023
14. Kostenko, G., Zaporozhets, A.: Transition from Electric Vehicles to Energy Storage: Review on Targeted Lithium-Ion Battery Diagnostics. **Energies**. 17(20), 5132 (2024). https://doi.org/10.3390/en17205132
15. GLPK (GNU Linear Programming Kit). Linear Programming and Algorithms for Communication Networks. 2012. pp. 39–44 (2023). https://doi.org/10.1201/b12733-5. Date of Access: 01 Aug 2023
16. Regulation on Approval of the Transmission System Code. Decision of the National Commission for State Regulation of Energy and Public Utilities No. 309 of March 14, 2018. Updated as of April 14, 2023. Available at: https://zakon.rada.gov.ua/laws/show/v0309874-18#Text. Accessed: 27 July 2023
17. Projected costs of generating electricity 2020—analysis. IEA. https://www.iea.org/reports/projected-costs-of-generating-electricity-2020. Date of Access: 21 July 2023
18. Central collection and publication of electricity generation, transportation and consumption data and information for pan-European market. ENTSO-E Transparency Platform. https://transparency.entsoe.eu/load-domain/r2/yearLoad/show. Date of Access: 21 July 2023
19. Central collection and publication of electricity generation, transportation and consumption data and information for pan-European market. ENTSO-E Transparency Platform. https://transparency.entsoe.eu/generation/
20. Balance of Flows Between the Energy System of Ukraine and the Energy Systems of Neighboring Countries. Official Website of NPC "Ukrenergo". https://ua.energy/peredacha-i-dyspetcheryzatsiya/dyspetcherska-informatsiya/peretoky/

Chapter 4
Modeling and Synchronizing Energy Systems of Ukraine and Europe: A 2050 Perspective

4.1 Calculation of Forecast Parameters for the Installed Capacity of Solar Power Plants (SPPs) for Ukraine and Germany Until 2050 Using the Dynamic Model of Cyclical Development of Generation Systems

The calculation of forecast parameters for the installed capacity of SPPs for Ukraine and Germany until 2050 was carried out using the methods presented in Chap. 2 ot this book:

- values of the coefficient of dynamics of economic and technological indicators (CDETIs) of economic and production development for Ukraine and Germany (Fig. 4.1) and in summarized Tables 4.1 and 4.2;
- functional of economic and technological influence (2.3);
- life cycle model of innovative technology in the form of a generalized logistic curve, which is a sum of three logistic curves *J1, J2, J3* (2.4).

Thus, using the economic-mathematical model of innovation diffusion in energy, scenarios for the development of solar energy systems in Germany and Ukraine have been calculated (Fig. 4.2).

A. Zaporozhets et al., *Structure Optimization of Power Systems with Renewable Energy Sources*, Studies in Systems, Decision and Control 583,
https://doi.org/10.1007/978-3-031-83697-8_4

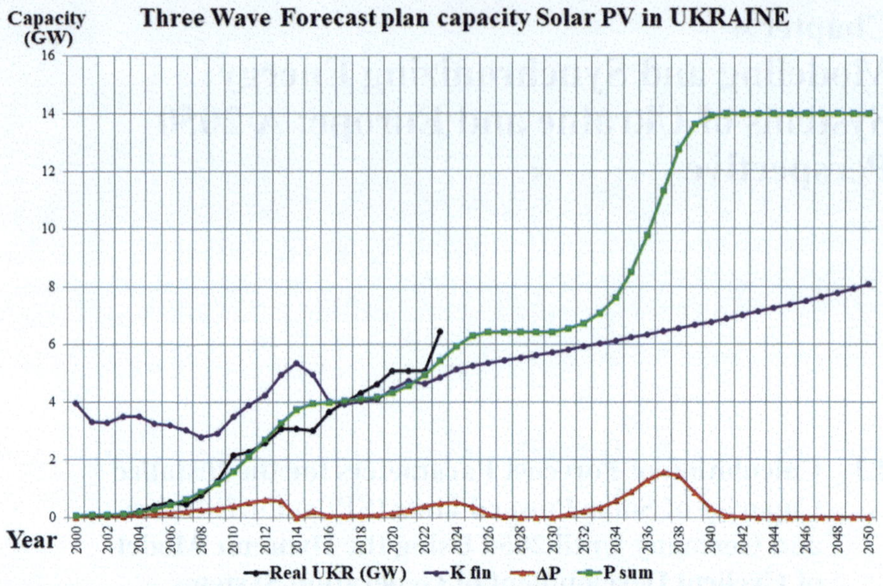

Fig. 4.1 Scenario of solar energy development in Ukraine until 2050

Table 4.1 Abbreviated table of forecasted parameters for installed capacity of PV systems for Ukraine

Year	Installed capacity, RealUKR (GW)	CDETI, K fin	*J1*	*J2*	*J3*	Forecast installed capacity P, (GW)	ΔP, (GW)
2000	0.03	4	0.0	0.0	0.0	0.1	0.0
2005	0.40	3	0.3	0.0	0.0	0.3	0.1
2010	2.14	3	1.6	0.0	0.0	1.6	0.4
2015	3.02	5	3.9	0.0	0.0	4.0	0.2
2020	5.06	4	4.0	0.3	0.0	4.3	0.1
2025			4.0	2.3	0.0	6.3	0.4
2030			4.0	2.4	0.0	6.4	0.0
2035			4.0	2.4	2.1	8.5	0.9
2040			4.0	2.4	7.5	13.9	0.3
2045			4.0	2.4	7.6	14.0	0.0
2050			4.0	2.4	7.6	14.0	0.0

Table 4.2 Abbreviated table of forecasted parameters for installed capacity of PV systems for Germany

Year	Installed capacity, RealUKR (GW)	CDETI, K fin	*J1*	*J2*	*J3*	Forecast installed capacity P, (GW)	ΔP, (GW)
2000	0	100	0	0	0	0	0
2005	2	86	2	0	0	2	4
2010	18	74	18	0	0	18	29
2015	40	64	39	0	0	39	3
2020	48	55	40	6	0	46	10
2025		48	40	29	0	69	16
2030		41	40	34	0	74	1
2035		35	40	34	6	80	8
2040		30	40	34	24	98	18
2045		25	40	34	40	114	8
2050		20	40	34	45	119	2

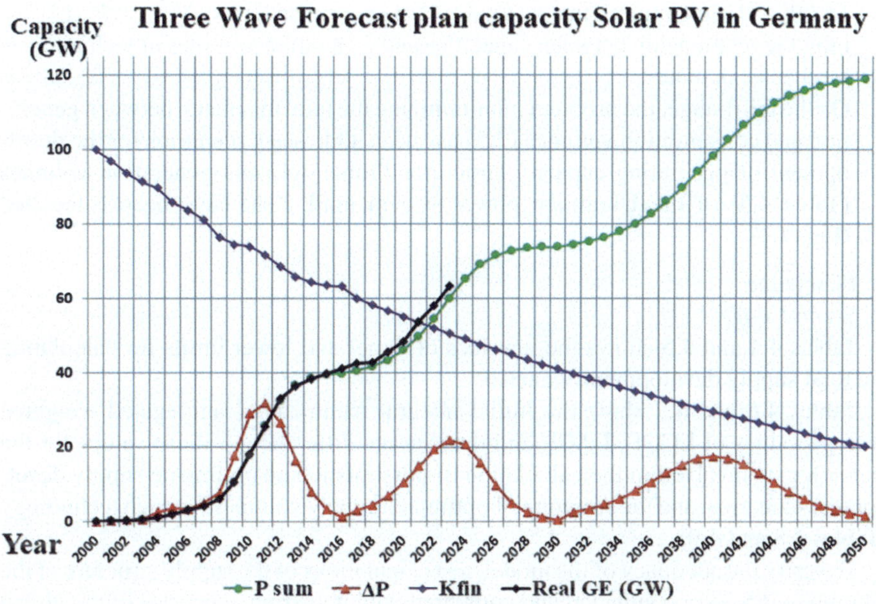

Fig. 4.2 Scenario of solar energy development in Germany until 2050

4.2 Calculating Forecasts for the Structure of Electricity Consumption and Generation in the IPS of Ukraine Until 2040 Using the Method of Economic-Technological Forecasting of Optimal Operation and Development Parameters of the Energy System, Along with the Long-Term Technological Renewal Model

According to the method of economic-technological forecasting of optimal operation and development parameters of the energy system, the forecast is calculated in three stages:

1. On the first stage, using the generalized logistic model (2.4) and formula (2.6), pessimistic and optimistic forecasts of electricity consumption volumes for the IPS of Ukraine until 2040 have been calculated (Table 4.3) and (Fig. 2.2).
2. On the second stage, using the formulas of net benefits $NB_k = K_m(LACE_k - LCOE_k)$ [1] and minimizing the total cost of generation $C_{S\tau}^{\Sigma} = \sum_{k=1}^{K} NB_k \cdot E_{S\tau}^{k} \rightarrow \min$, the relationship between capacities and generation volumes of technologies has been calculated.
3. On the third stage, the problem of minimizing the total mismatch between generation and consumption volumes (2.5) has been addressed, scenarios for the development of generating capacity structures (Table 4.4) and generation volumes (Table 4.5) of the Ukrainian power system until 2040 have been calculated (Fig. 4.3).

First stage

Second stage.

Tables 4.4 and 4.5 shows the matrices of upper and lower limits for calculating the total supply development scenario.

Tables 4.6 and 4.7 show the initial integral parameters—aggregated weighted average values of LCOE, LACE and the average difference of these values for the technologies involved in the calculation of pessimistic and optimistic supply development scenarios, and the structure of optimized supply volumes for each technology (Tables 4.8 and 4.9).

To verify the adequacy of the model, test calculations of the supply structure of the Ukrainian IPS were conducted and compared with numerous scenarios of the global energy supply structure presented by various researchers [2, page 4]. The results of the comparative test calculations of the supply structure of the Ukrainian IPS and global energy supply are presented in Fig. 4.3.

The obtained results allow us to conclude sufficient alignment between the compared scenarios, confirming the adequacy of the proposed model and its potential for developing scenarios for the development and updating of energy systems.

Thus, the long-term technological updating model of the energy system structure enabled forecasts of electricity consumption and generation structures in Ukraine's

Table 4.3 Pessimistic and optimistic forecasts of electricity consumption for the Unified Energy System of Ukraine until 2040, billion kWh

Year	Actual consumption	Total consumption pessimistic	Total consumption optimistic
2015	148	148	148
2016	151	152	152
2017	154	153	153
2018	158	153	153
2019	152	153	153
2020	146	152	152
2021	153	149	149
2022	144	145	145
2023		141	141
2024		140	140
2025		140	140
2026		151	156
2027		162	172
2028		172	187
2029		181	200
2030		190	213
2031		197	223
2032		202	232
2033		206	239
2034		210	245
2035		212	249
2036		214	252
2037		215	254
2038		216	256
2039		217	257
2040		217	257

IPS until 2040, allowing assessment of the potential for enhancing components of Ukraine's IPS.

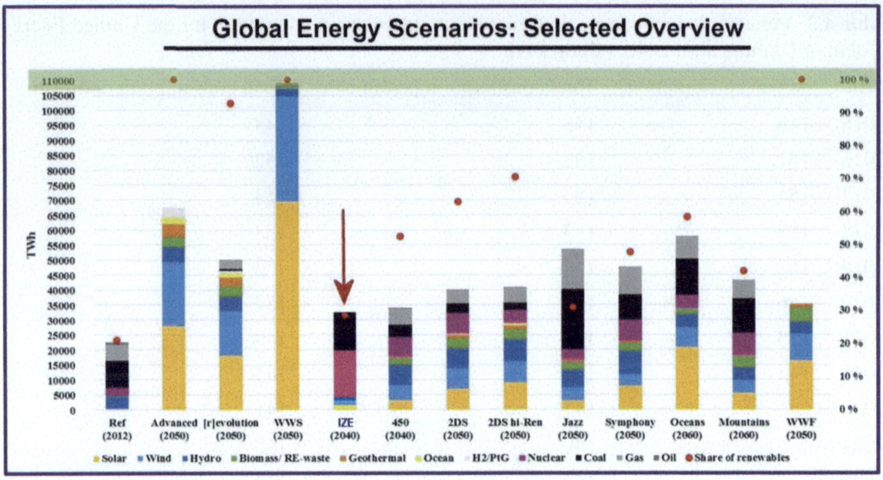

Fig. 4.3 Results of comparative test calculations of the supply structure of the Ukrainian IPS and global energy supply

Table 4.4 Matrices of upper and lower limits for calculating the pessimistic scenario of the development of total supply, billion kWh

	Lower limits					Upper limits				
	2020	2025	2030	2035	2040	2020	2025	2030	2035	2040
TPP	54	29	22	30	38	54	86	65	87	110
HPP	2.2	2.4	2.7	3	3.3	2.2	6.1	6.7	7.3	8
NPP	76	52	50	51	51	76	125	120	121	123
WPP	0.25	1.9	1.6	3.2	4	0.7	5.7	4.6	9.5	12
SPP	0.13	3.3	3.1	4	4.2	0.4	9.6	9.2	12	12.4

Table 4.5 Matrices of upper and lower limits for calculating the optimistic scenario of the development of total supply, billion kWh

	Lower limits					Upper limits				
	2020	2025	2030	2035	2040	2020	2025	2030	2035	2040
TPP	54	29	22	30	38	54	86	65	87	110
HPP	2.2	2.4	2.7	3	3.3	2.2	6.1	6.7	7.3	8
NPP	76	52	50	51	51	76	125	128	134	134
WPP	0.25	1.9	1.6	3.2	4	0.7	5.7	4.6	9.5	12
SPP	0.13	3.3	3.1	4	4.2	0.4	9.6	9.2	12	12.4

Table 4.6 Initial integrated parameters of supply technologies and structure of optimized supply volumes for the pessimistic scenario

Aggregated weighted average LCOE and LACE for 2040, (2016$/MWh)								
Technology	LCOE	LACE	Difference	Volumes of supply that are being optimized				
Dispatchable technologies				2020	2025	2030	2035	2040
TPP	116.55	66.8	– 49.85	54	35.6	64	81	74
HPP	62.4	66.1	6.95	5.7	2.5	6.2	6.5	8
NPP	89.6	65.8	– 23.8	76	96	106	105	111
Total optimized supply				136	134	176	193	193
Technologies that cannot be dispatched								
WPP	57.6	60.9	3.3	0.7	2	4.2	8.3	12
SPP	63.9	72.2	8.2	0.4	4	9.1	11.3	12
Total non-optimized supply				1.1	6	13.3	19.6	24
Supply vector, billion kWh				136.6	140	190	212	217
Target supply, billion kWh				152	140	190	212	217
Supply volume difference				15.3	0	0	0	0

Table 4.7 Initial integrated parameters of supply technologies and structure of optimized supply volumes for the optimistic scenario

Aggregated weighted average LCOE and LACE for 2040, (2016$/MWh)								
Technology	LCOE	LACE	Difference	Volumes of supply that are being optimized				
Dispatchable technologies				2020	2025	2030	2035	2040
TPP	116.55	66.8	– 49.85	54	36	65	87	102
HPP	62.4	66.1	6.95	5.7	2.5	6.7	7.3	8
NPP	89.6	65.8	– 23.8	76	96	128	133	123
Total optimized supply				136	134	199	227	233
Technologies that cannot be dispatched								
WPP	57.6	60.9	3.3	0.7	2	4.6	9.5	12
SPP	63.9	72.2	8.2	0.4	4	9.2	12	12.3
Total non-optimized supply				1.1	6	13.8	21.5	24.3
Supply vector, billion kWh				136.6	140	213	249	258
Target supply, billion kWh				152	140	213	249	258
Supply volume difference				15.3	0	0	0	0

Table 4.8 Scenario for the development of the structure of generating capacities of the IPS of Ukraine until 2040

	Generation	2020 Actual	2025	2030	2035	2040
	Egen, billion (kWh)	Pinst, GW	Pinst, ОПТИМ, GW	Pinst, ОПТИМ, GW	Pinst, ОПТИМ, GW	Pinst, ОПТИМ, GW
Total	143.7	52.4	42.1	72.4	90.7	99.3
TPP	53.9	28.0	19.8	35.9	48.3	56.6
HPP	76.0	13.8	16.0	21.2	22.1	20.4
NPP	5.7	4.8	1.8	4.8	5.3	5.8
WPP	3.1	1.0	0.8	1.8	3.7	4.6
SPP	5.0	4.8	3.8	8.8	11.3	11.8

Table 4.9 Scenario of the development of the structure of IPS generation volumes of Ukraine until 2040

	Generation	2020 Actual	2025	2030	2035	2040
	Egen, billion (kWh)	Egen, billion (kWh)	Egen, billion (kWh)	Egen, billion (kWh)	Egen, billion (kWh)	Egen, billion (kWh)
Total	143.7	136.6	140.4	212.9	248.8	257.5
TPP	53.9	53.9	35.7	64.7	87.2	102.2
HPP	76.0	76.0	96.2	127.8	132.9	123.0
NPP	5.7	5.7	2.5	6.7	7.3	8.0
WPP	3.1	0.7	2.0	4.6	9.5	11.9
SPP	5.0	0.4	4.0	9.2	11.9	12.4

4.3 Modeling the Parallel Operation of the IPS of Ukraine with Neighboring Countries Included in ENTSO-E Using the Model of a Multi-Nodal Integrated Energy System

The relevance of modeling parallel operation of Ukraine's power system with neighboring countries within ENTSO-E is driven by several key factors:

- International climate agendas continually raise requirements for the environmental sustainability of energy systems.
- Integration of a significant share of intermittent renewable energy sources (RES), particularly WPP and SPP, into energy systems:

  a. Complicates the planning processes of energy system operations even in the short-term perspective (for the next day and week), particularly in the distribution of active power reserves among power plants (thermal, combined heat and power, hydroelectric, and pumped storage).

b. Requires the implementation of new types of industrial-scale capacities (such as battery storage, demand response technologies, "maneuverable" nuclear power plants, electric boilers, hydrogen energy technologies, etc.) capable of enhancing the overall flexibility of the energy system and ensuring guaranteed support for specified frequency parameters.

- The national energy systems of developed countries are undergoing transformation both in terms of their generating capacity structures and the principles governing energy market regulation.

The introduction of advanced electronics into energy systems (smart-metering) allows for significant reduction in information processing time, formation and transmission of control signals, and enables management of electricity consumption patterns by consumers (prosumers, electric vehicles, etc.). This capability provides greater flexibility in energy system planning but requires much more sophisticated modeling tools.

4.4 Conditions for the Parallel Operation of the Ukrainian Integrated Power System with Neighboring Countries Connected to ENTSO-E

The large-scale reform of relations in the electricity sector in recent years in Ukraine is driven by the desire for integration into the European Union, which is enshrined at the legislative level, specifically by the Law of Ukraine "On the Ratification of the Agreement between Ukraine and the European Union" [3]. One of the important steps towards Ukraine's integration into the EU is the connection of Ukraine's Integrated Power System (IPS) to the association of Continental Europe transmission system operators (ENTSO-E). For this purpose, in July 2019, Ukraine implemented a new electricity market model built on EU-compatible principles, which currently operates according to the provisions of the Law of Ukraine "On the Electricity Market of Ukraine" [4] and subordinate regulatory acts, including the "Market Rules" [5], the "Rules of the Day-Ahead Market" and the "Intraday Market" [6].

One of the key conditions for the integration of Ukraine's IPS with the Continental Europe transmission system operators' association (ENTSO-E) is the balanced self-sufficiency of the national energy system. Given the current level of installed renewable energy capacities, this requires additional consideration of the natural stochasticity of wind and solar power generation, the probabilistic nature of accidental failures of energy system components, and the search for additional opportunities to ensure the balance of the energy system. These opportunities can be realized by introducing new dispatch modes for units covering reserves and energy storage. If these measures are insufficient, additional flexible storage and reserve power generation capacities will be required [7–9]. Another measure to ensure the stable and balanced operation of Ukraine's IPS is the integration of Ukraine's energy system with the Continental European energy system ENTSO-E [10, 11]. In this regard, on September 12, 2019, Ukrenergo signed a "Service Contract with the Consortium of System Operators of ENTSO-E for the Study of the Possibility of Synchronous Integration of Ukraine's Energy System with ENTSO-E.

According to the "Plan for the Integration of the Ukrainian IPS into ENTSO-E," the Ukrainian energy system will be synchronized with the continental European network alongside Moldova's energy system. The Ukrainian Transmission System Operator, "Ukrenergo", signed the Agreement on the Conditions for Future Synchronization on June 28, 2017. The start of the synchronous operation of the Ukrainian energy system with the continental European network was initially scheduled for 2023 but was completed in 2022 [12]. The total cost of integration measures with ENTSO-E is estimated at 11.4 billion UAH, of which 4327 million UAH will be financed by international financial organizations (Table 4.10). The expected direct annual economic benefit after synchronization, assuming free cross-border electricity exchange within the capacity of 4 GW, is more than 40 billion UAH [12].

As part of the implementation of the mentioned plan [12], in June 2020, the government, Ukrenergo, and the World Bank decided to build a direct current insertion. In the second quarter of 2021, additional static calculations and studies on the conditions for the synchronous connection of the continental European and Ukrainian/Moldovan energy systems were planned, as well as the necessary dynamic system studies. In 2022, the test operation of the Ukrainian IPS and the energy system of the Republic of Moldova in isolated mode was planned and carried out. In the first quarter of 2023, the Ukrainian IPS was scheduled to begin synchronous operation with the Continental European ENTSO-E energy system, which was achieved in March 2022 [13]. Additionally, the service contract envisaged developing a general mathematical model of the Ukrainian energy system and ENTSO-E, as well as studying the dynamic and static stability of the energy system. These studies are an integral part of the Agreement on the conditions for the future integration of the Ukrainian IPS into ENTSO-E.

Table 4.10 Cost of measures at the expense of the MFI

Million hryvnias	List of expenditure
4114	Transmission Network
4048	Measures to ensure generation readiness for operation within ENTSO-E
2965	Organization of communication channels for technological management of the Ukrainian IPS
157	Conducting additional studies on static and dynamic stability
75.8	Compliance with ENTSO-E Operational Handbook requirements
11.7	Operation of the Ukrainian IPS in "island" mode
5	Certification of the TSO

European countries differ significantly in terms of electricity export–import: the largest exporters are Germany and France; exports exceed imports in Switzerland and the Czech Republic, as well as in Norway and Sweden. The largest importer is Italy, which depends on electricity supplies not only from its immediate neighbors but also from Germany. The second largest importer is the United Kingdom, which receives electricity from France. The highest percentage of imports is characteristic of "small" countries (Finland, Austria, Hungary). In countries with large volumes of domestic consumption, the share of imports is small.

The presence of inter-state power lines between Ukraine's Integrated Power System (IPS) and neighboring Transmission System Operators (TSOs) of continental Europe allows for the optimization of structure and modes, as well as mutually beneficial electricity exchange with the energy systems of European Union countries (Hungary, Slovakia, Poland, and Romania). The existence of cross-border electricity exchange with neighboring TSOs of continental Europe adjacent to Ukraine's IPS is partly due to the specific structure of generating capacities in each country (Table 4.7). For example, in Ukraine, the installed capacity of nuclear power plants (NPPs) is 13,835 MW, while Poland has no NPPs at all, with the largest generation coming from thermal power plants (TPPs), which have an installed capacity of 28,794 MW (Table 4.11).

The IPS of Ukraine is divided into two parts. The main part of the IPS of Ukraine operates in parallel with the energy union of the Commonwealth of Independent States (CIS) and Baltic countries, while the other part, the "Burshtyn TPP Island," operates within the Continental European electricity system ENTSO-E

Table 4.11 Installed Generation Capacity Values (MW) of the Energy Systems of Ukraine and Neighboring European Union Countries (Hungary, Poland, and Romania, unfortunately, such data are not available for Slovakia) as of early 2021

Generating technology	Ukraine	Poland	Romania	Hungary
NPP	13,835		1300	1899
TPP of the GK by type of fuel:	21,842	28,794		5498
Coal of the gas group	10,723	272		
Coal of the anthracite group	6519	20,925	986	42
Brown coal		7597	2483	1007
Gas/fuel oil	4600	359	2184	4029/420
CHP	6104.5	2555		
HPP	4829	604	6134	58
PSPP	1487.8	1794		
SPP	5391.6	3473	1145	1458
WPP	1111.2	6570	2957	323
Biofuel stations	199.5	880	117	241
Other RES				73
Together	54,800.6	45,029	17,306	9612

(Table 4.12) [10]. This has enhanced the reliability of its operation and reduced the capacity of the total required emergency reserve. Ukrenergo transmits electricity via interstate transmission lines from the "Burshtyn TPP Island" to Hungary, Slovakia, and Romania. Currently, the maximum allowable export capacity is up to 650 MW. According to [14], the forecasted interconnector capacities are as follows: Poland—705 MW, Hungary—235 MW, Slovakia—210 MW, Romania—150 MW. Each neighboring country is characterized by [10] relatively different Schedule of Electrical Loads (SEL), different degrees of utilization of their capacities, and shifts in consumption peaks/minimums. For instance, the peak daily load period in the IPS of Ukraine in 2018 occurred on December 20. The actual values of consumption parameters, coverage, and distribution of generating technologies that ensured the coverage of the electrical load schedules [15–17] on December 20, 2018, are presented in Table 4.13.

Comparative SEL of actual electrical load of Ukraine and neighboring countries—Poland, Hungary, Romania, and Slovakia—for the peak days of December 20 and the minimum days of May 27 of the daily load in the IPS of Ukraine in 2018 (Kyiv time) [15–17] are presented in Table 4.14.

Comparative SEL of actual electrical load of Ukraine and neighboring countries—Poland, Hungary, Romania, and Slovakia—for the peak days of February 28 and the minimum days of May 20 of the daily load in the UES of Poland in 2018 (Kyiv time) [15–17] are presented in Table 4.15.

Comparative SEL of actual electrical load of Ukraine and neighboring countries—Poland, Hungary, Romania, and Slovakia—for the peak days of March 2 and the minimum days of May 21 of the daily load in the UES of Hungary in 2018 (Kyiv time) [15–17] are presented in Table 4.16.

Comparative SEL of actual electrical load of Ukraine and neighboring countries—Poland, Hungary, Romania, and Slovakia—for the peak days of February 27 and the minimum days of April 8 of the daily load in the UES of Romania in 2018 (Kyiv time) [15–17] are presented in Table 4.17.

Comparative SEL of actual electrical load of Ukraine and neighboring countries—Poland, Hungary, Romania, and Slovakia—for the peak days of February 27 and the minimum days of July 29 of the daily load in the **UES of Slovakia** in 2018 (Kyiv time) [15–17] are presented in Table 4.18.

As shown in Table 4.12, the highest volumes of electricity exports to Hungary, Romania, and Poland occurred in the following months of 2020:

- January—697 million kWh;
- February—740 million kWh;
- March—774 million kWh.

In contrast, the volumes of imports in the same months were 561 million kWh, 551 million kWh, and 443 million kWh respectively from Slovakia, Hungary, and Belarus.

The analysis conducted in [18] of the current state of generating capacities in the Ukrainian power system until February 24, 2022, indicates that Ukraine currently

Table 4.12 Actual data on the volumes of exported/imported electricity (million kWh) from/to IPS of Ukraine in 2020

Country	January	February	March	April	May	June	July	August	September	October	November	December	Total
Export													
Hungary	278	372	423	129	0	0	4	17	269	236	213	170	2110
Slovakia	0	0	0	0	0	0	2	1	12	1	3	22	41
Romania	223	177	149	21	0	0	1	6	43	97	130	23	868
Poland	151	149	159	88	125	102	114	113	108	149	123	103	1484
Moldova	45	43	42	0	0	0	0	0	0	15	11	11	167
Russia	0	0	0	0	0	0	0	0	0	0	0	0	0
Belarus	0	0	0	0	0	0	12	35	36	0	0	0	83
Import													
Hungary	97	178	97	37	16	9	11	19	9	23	11	9	516
Slovakia	329	270	309	103	14	8	10	18	44	105	145	118	1474
Romania	24	20	27	8	2	2	1	2	1	0	1	1	89
Poland	0	0	0	0	0	0	0	0	0	0	0	0	0
Moldova	0	0	0	0	0	0	0	0	0	0	0	0	0
Belarus	93	49	8	3	0	0	0	0	0	0	0	0	152
Russia	18	34	1	0	0	0	0	0	0	0	0	0	53
Total													
Export	697	740	774	238	125	102	133	171	468	497	479	0	4754
Import	561	551	443	151	33	19	22	39	54	129	157	0	2285
The volumes of technological flows of electricity are determined by the parallel operation of the IPS of Ukraine and neighboring countries													
Russia	3.5	4.6	6.4	4.7	5.2	5.3	4.5	5.6	3.6	5.9	5.2	4.5	58.9
Belarus	0.1	0.1	0.1	0.1	0.1	0.1	0.1	0.1	0.1	0.1	0.1	0.0	0.9

Table 4.13 Actual values of parameters of consumption, coverage and distribution of capacities (MW) of generating technologies that ensured coverage of SEL during the period of maximum daily load in **IPS of Ukraine** on December 20, 2018

Hours	Consumption	Load	NPP	TPP of the GK	CHP	HPP	Small HPP	PSPP gen	PSPP pump	SPP	WPP
1	18,872	20,433	11,000	6276	2335	991	19	0	– 939	0	317
2	18,316	20,135	10,980	6415	2323	181	19	0	– 938	0	317
3	18,192	20,169	10,974	6348	2273	574	18	0	– 1377	0	278
4	18,212	20,439	10,995	6231	2278	1366	18	0	– 1145	0	243
5	18,457	20,374	10,979	6365	2282	990	18	0	– 1147	0	244
6	19,205	19,869	10,964	6195	2356	198	18	0	0	0	246
7	20,721	21,400	10,997	7141	2394	1217	17	0	0	9	242
8	21,798	22,527	10,976	7786	2366	2282	18	0	0	17	232
9	22,899	23,502	10,985	8464	2319	2244	28	323	0	28	247
10	23,403	24,094	10,968	8776	2330	2199	25	645	0	43	220
11	23,577	24,250	10,989	9139	2329	1791	25	644	0	55	186
12	23,355	24,047	11,003	9203	2336	1246	24	644	0	64	162
13	23,205	23,912	11,014	9298	2334	1422	24	324	0	63	156
14	23,318	24,052	10,996	9302	2330	1730	24	323	0	46	178
15	23,127	23,980	10,979	9315	2323	1686	24	326	0	33	149
16	23,488	24,157	10,976	9307	2361	2388	24	151	0	0	156
17	23,667	24,603	10,988	9142	2359	2575	25	661	0	0	153
18	23,344	24,148	11,003	8765	2358	2378	26	660	0	0	160
19	22,901	23,445	10,982	8615	2346	2371	25	187	0	0	117
20	22,587	23,299	11,005	8592	2370	2419	25	0	0	0	110
21	22,147	22,859	10,987	8597	2373	1602	24	0	0	0	89
22	21,768	22,647	10,987	8553	2369	629	15	318	0	0	98
23	20,513	21,251	10,993	7480	2374	620	14	0	0	0	87
24	19,881	21,462	10,994	7724	2375	567	13	0	-948	0	79

has sufficient generating capacities, even considering that a portion of these capacities is located in non-controlled territories. However, the majority of these capacities are base-load and are not designed for frequent and rapid changes in operational modes. Additionally, the portion of capacities that can adjust their operational modes (primarily concentrated in TPPs) has already reached the end of their operational lifespan. Therefore, the Ukrainian power system is characterized as inflexible, meaning it lacks adequate maneuvering capacities.

Based on the analysis presented in [18], scenarios for demand growth and the structure of generating capacities have been proposed. Depending on the peak load and levels of electricity consumption, it is noted that the required capacity for SPPs

Table 4.14 Comparative SEL of actual electrical load of Ukraine and neighboring countries—Poland, Hungary, Romania, and Slovakia—for the peak days of December 20 and the minimum days of May 27 of the daily load in the **IPS of Ukraine** in 2018 (Kyiv time)

Hour	The day of maximum load 20 December 2018					The day of minimum load 27 May 2018				
	Ukraine	Poland	Hungary	Romania	Slovakia	Ukraine	Poland	Hungary	Romania	Slovakia
1	20,433	20,082	5224	6994	3371	13,658	14,684	4023	5616	2721
2	20,135	18,893	4972	6880	3282	13,659	13,969	3864	5332	2615
3	20,169	18,349	4738	6845	3284	13,341	13,627	3639	5178	2547
4	20,439	18,042	4614	6934	3286	13,243	13,412	3552	5073	2551
5	20,374	18,005	4617	7184	3342	13,201	12,876	3512	5056	2527
6	19,869	18,305	4860	7730	3527	13,050	12,639	3429	4910	2491
7	21,400	19,103	5492	8610	3938	13,444	13,386	3558	4862	2603
8	22,527	21,768	6072	9139	4146	13,664	14,390	3862	5062	2790
9	23,502	23,792	6149	9319	4254	13,762	15,640	4106	5297	3006
10	24,094	24,593	6221	9359	4327	14,312	16,637	4370	5465	3197
11	24,250	24,835	6270	9272	4228	14,300	17,208	4540	5527	3296
12	24,047	24,778	6246	9140	4293	14,473	17,328	4624	5494	3316
13	23,912	25,022	6321	9073	4227	14,383	17,390	4657	5538	3218
14	24,052	25,063	6263	8968	4202	14,376	17,264	4548	5534	3106
15	23,980	25,322	6174	8947	4155	14,223	17,034	4497	5477	3086
16	24,157	24,902	6209	9081	4104	14,335	16,705	4513	5449	3054
17	24,603	25,070	6304	9494	4252	14,430	16,524	4493	5501	3054
18	24,148	25,691	6338	9443	4216	14,301	16,502	4472	5548	3026
19	23,445	25,517	6217	9300	4154	14,564	16,669	4516	5624	3043
20	23,299	25,187	6141	9131	4078	14,244	16,920	4561	5723	3086

(continued)

Table 4.14 (continued)

Hour	The day of maximum load 20 December 2018					The day of minimum load 27 May 2018				
	Ukraine	Poland	Hungary	Romania	Slovakia	Ukraine	Poland	Hungary	Romania	Slovakia
21	22,859	25,061	5927	8690	4011	14,669	17,124	4625	5824	3168
22	22,647	24,507	5781	8121	3771	15,277	17,665	4851	6198	3155
23	21,251	23,013	5569	7595	3597	14,011	17,145	4628	6044	3008
24	21,462	21,382	5371	7223	3444	13,785	15,804	4294	5618	2859

Table 4.15 Comparative SEL of actual electrical load of Ukraine and neighboring countries—Poland, Hungary, Romania, and Slovakia—for the peak days of February 28 and the minimum days of May 20 of the daily load in the **UES of Poland** in 2018 (Kyiv time)

	The day of maximum load 28 February 2018					The day of minimum load 20 May 2018				
Hour	Ukraine	Poland	Hungary	Romania	Slovakia	Ukraine	Poland	Hungary	Romania	Slovakia
1	20,719	20,025	5333	7457	3646	13,723	14,271	3819	5463	2735
2	20,263	19,365	5049	7271	3551	14,040	13,588	3659	5287	2618
3	20,431	19,063	4840	7183	3527	14,113	13,183	3468	5118	2554
4	20,392	19,106	4731	7164	3523	13,990	13,031	3378	5004	2549
5	20,232	19,474	4762	7273	3637	14,062	12,551	3320	5045	2536
6	20,056	20,350	5058	7457	3787	13,971	12,262	3207	4966	2499
7	21,539	22,455	5675	7969	4167	14,103	12,757	3272	4932	2638
8	22,449	24,431	6184	8661	4336	13,840	13,501	3522	5095	2796
9	23,528	25,582	6252	9273	4409	14,498	14,591	3729	5393	2975
10	23,922	25,806	6306	9511	4432	15,068	15,379	3936	5554	3171
11	23,937	25,604	6297	9509	4332	15,148	15,690	4017	5583	3277
12	23,817	25,759	6350	9366	4356	15,447	15,698	4023	5548	3308
13	23,739	25,766	6322	9453	4304	15,462	15,640	4035	5592	3183
14	23,500	25,700	6296	9297	4276	15,484	15,538	3955	5558	3064
15	23,246	25,387	6191	9011	4248	15,303	15,327	3897	5421	2999
16	23,420	24,903	6227	8842	4196	15,281	14,930	3864	5402	2944
17	23,041	24,737	6161	8820	4241	15,433	14,762	3874	5438	2962
18	23,784	25,435	6192	8894	4323	15,398	14,749	3896	5520	2941
19	23,859	26,258	6497	9385	4413	15,633	15,104	3904	5613	2962
20	23,469	26,081	6537	9606	4381	15,645	15,473	3951	5732	3026

(continued)

Table 4.15 (continued)

Hour	The day of maximum load 28 February 2018					The day of minimum load 20 May 2018				
	Ukraine	Poland	Hungary	Romania	Slovakia	Ukraine	Poland	Hungary	Romania	Slovakia
21	23,030	25,439	6374	9449	4291	16,165	15,941	4010	6011	3120
22	22,273	23,893	6183	9037	4053	16,394	16,567	4310	6410	3121
23	21,243	22,307	5875	8389	3843	15,033	15,940	4149	6207	2991
24	21,070	20,934	5686	7823	3705	14,159	14,847	3906	5637	2834

Table 4.16 Comparative SEL of actual electrical load of Ukraine and neighboring countries—Poland, Hungary, Romania, and Slovakia—for the peak days of March 2 and the minimum days of May 21 of the daily load in the **UES of Hungary** in 2018 (Kyiv time)

	The day of maximum load 2 March 2018					The day of minimum load 21 May 2018				
Hour	Ukraine	Poland	Hungary	Romania	Slovakia	Ukraine	Poland	Hungary	Romania	Slovakia
1	20,343	20,108	5347	7475	3550	14,277	13,948	3576	5219	2729
2	19,986	19,480	5112	7258	3463	13,842	13,460	3433	5003	2624
3	20,062	19,235	4891	7127	3454	14,075	13,283	3250	4916	2590
4	20,281	19,149	4756	7105	3459	14,097	13,221	3142	4884	2598
5	20,017	19,506	4767	7151	3529	13,888	13,151	3096	4925	2638
6	19,878	20,342	5036	7329	3677	14,181	13,799	3016	5028	2768
7	21,171	22,365	5672	7840	4051	14,975	16,878	3132	5425	3177
8	22,269	24,385	6211	8585	4269	15,590	19,323	3371	6181	3365
9	23,272	25,385	6330	9152	4341	15,999	20,555	3591	6592	3501
10	23,557	25,585	6435	9392	4449	16,110	21,077	3840	6681	3520
11	23,576	25,473	6500	9315	4403	16,107	20,906	3927	6732	3491
12	23,531	25,655	6547	9276	4508	16,423	21,178	3986	6709	3525
13	23,342	25,484	6568	9144	4422	16,334	21,400	4019	6704	3585
14	23,417	25,578	6531	9057	4424	16,569	21,398	3946	6774	3556
15	23,307	25,071	6400	8881	4359	16,615	20,907	3911	6713	3546
16	22,928	24,593	6368	8774	4284	16,363	20,742	3921	6627	3473
17	23,013	24,448	6291	8731	4270	16,308	20,463	3950	6593	3467
18	23,452	25,075	6230	8844	4267	16,187	20,156	3980	6529	3411
19	23,521	25,907	6393	9223	4314	16,155	19,841	4040	6503	3376
20	23,060	25,764	6358	9326	4242	15,956	20,136	4111	6616	3434

(continued)

Table 4.16 (continued)

	The day of maximum load 2 March 2018					The day of minimum load 21 May 2018				
Hour	Ukraine	Poland	Hungary	Romania	Slovakia	Ukraine	Poland	Hungary	Romania	Slovakia
21	22,805	24,917	6149	9125	4132	16,653	20,352	4245	6809	3482
22	21,781	23,381	5979	8727	3901	16,755	20,238	4418	7147	3407
23	21,222	22,016	5710	8086	3699	15,468	18,684	4184	6721	3179
24	20,995	20,843	5523	7557	3570	14,990	17,220	3880	6173	3010

Table 4.17 Comparative SEL of actual electrical load of Ukraine and neighboring countries—Poland, Hungary, Romania, and Slovakia—for the peak days of February 27 and the minimum days of April 8 of the daily load in the **UES of Romania** in 2018 (Kyiv time)

	The day of maximum load 27 February 2018					The day of minimum load 8 April 2018				
Hour	Ukraine	Poland	Hungary	Romania	Slovakia	Ukraine	Poland	Hungary	Romania	Slovakia
1	19,976	19,726	5205	7529	3580	14,685	15,218	4218	5473	2800
2	19,998	19,238	4995	7316	3469	14,580	14,548	4039	5320	2681
3	20,301	18,970	4769	7247	3457	14,539	14,197	3850	5115	2660
4	20,207	19,037	4674	7232	3464	14,459	13,999	3770	5039	2649
5	20,168	19,261	4695	7309	3543	14,764	14,095	3707	4988	2653
6	20,009	20,066	4991	7472	3716	14,646	13,824	3794	4941	2664
7	21,153	22,166	5621	7946	4126	14,708	13,698	3804	4877	2766
8	22,321	24,179	6125	8684	4338	14,507	14,434	4124	4863	2910
9	23,503	25,370	6204	9259	4433	14,402	15,463	4372	4986	3104
10	24,187	25,600	6297	9529	4519	14,451	16,215	4549	5018	3260
11	24,254	25,513	6363	9596	4429	14,650	16,351	4576	4964	3340
12	24,072	25,766	6390	9525	4480	14,409	16,356	4594	4859	3346
13	23,928	25,791	6443	9453	4438	14,430	16,398	4537	4760	3197
14	23,918	25,938	6424	9376	4410	14,314	16,196	4395	4650	3118
15	24,046	25,530	6264	9128	4380	13,952	15,841	4264	4542	3039
16	23,787	25,050	6296	8940	4319	13,885	15,479	4200	4548	2980
17	23,585	24,778	6222	8962	4362	13,902	15,388	4199	4553	2987
18	24,005	25,572	6223	9021	4392	13,750	15,576	4249	4642	2990
19	24,280	26,221	6469	9526	4466	13,928	15,889	4299	4764	3053
20	23,937	26,174	6478	9695	4433	15,203	17,022	4514	5033	3191

(continued)

Table 4.17 (continued)

Hour	The day of maximum load 27 February 2018					The day of minimum load 8 April 2018				
	Ukraine	Poland	Hungary	Romania	Slovakia	Ukraine	Poland	Hungary	Romania	Slovakia
21	23,362	25,388	6271	9469	4298	16,389	18,199	5017	5748	3341
22	22,710	23,902	6125	9050	4084	15,652	17,415	4823	5881	3185
23	21,496	22,374	5787	8338	3915	15,256	16,506	4549	5579	3044
24	21,531	21,106	5589	7779	3731	14,270	15,456	4322	5251	2906

Table 4.18 Comparative SEL of actual electrical load of Ukraine and neighboring countries—Poland, Hungary, Romania, and Slovakia—for the peak days of February 27 and the minimum days of July 29 of the daily load in the **UES of Slovakia** in 2018 (Kyiv time)

	The day of maximum load 27 February 2018					The day of minimum load 29 July 2018				
Hour	Ukraine	Poland	Hungary	Romania	Slovakia	Ukraine	Poland	Hungary	Romania	Slovakia
1	19,976	19,726	5205	7529	3580	13,997	15,366	3986	5808	2661
2	19,998	19,238	4995	7316	3469	14,220	14,659	3784	5490	2541
3	20,301	18,970	4769	7247	3457	14,431	14,217	3615	5344	2498
4	20,207	19,037	4674	7232	3464	14,539	14,112	3530	5247	2450
5	20,168	19,261	4695	7309	3543	14,477	13,653	3469	5201	2463
6	20,009	20,066	4991	7472	3716	14,505	13,131	3424	5170	2395
7	21,153	22,166	5621	7946	4126	14,631	13,803	3503	5142	2516
8	22,321	24,179	6125	8684	4338	14,045	14,748	3803	5321	2687
9	23,503	25,370	6204	9259	4433	14,928	15,967	4087	5635	2891
10	24,187	25,600	6297	9529	4519	15,400	16,949	4361	5814	3098
11	24,254	25,513	6363	9596	4429	15,827	17,645	4562	5965	3229
12	24,072	25,766	6390	9525	4480	16,135	17,870	4675	6035	3259
13	23,928	25,791	6443	9453	4438	16,100	17,843	4749	6201	3179
14	23,918	25,938	6424	9376	4410	16,102	17,721	4678	6235	3111
15	24,046	25,530	6264	9128	4380	16,088	17,380	4640	6099	3022
16	23,787	25,050	6296	8940	4319	16,081	17,059	4641	6051	2996
17	23,585	24,778	6222	8962	4362	16,232	16,814	4656	6072	3027
18	24,005	25,572	6223	9021	4392	16,108	16,785	4648	6131	2989
19	24,280	26,221	6469	9526	4466	16,281	16,987	4680	6136	2993
20	23,937	26,174	6478	9695	4433	15,932	17,101	4718	6233	2998

(continued)

Table 4.18 (continued)

	The day of maximum load 27 February 2018					The day of minimum load 29 July 2018				
Hour	Ukraine	Poland	Hungary	Romania	Slovakia	Ukraine	Poland	Hungary	Romania	Slovakia
21	23,362	25,388	6271	9469	4298	16,548	17,196	4748	6402	3055
22	22,710	23,902	6125	9050	4084	16,806	17,609	4999	6669	3076
23	21,496	22,374	5787	8338	3915	15,395	17,051	4727	6353	2942
24	21,531	21,106	5589	7779	3731	14,649	15,830	4328	5907	2774

ranges from 6 to 7 GW, and for WPPs from 2.4 to 4 GW to meet the production volumes of electricity as outlined in Ukraine's New Energy Strategy (NES) until 2035: safety, energy efficiency, competitiveness [19].

When assessing electricity demand in the perspective of the next 10–15 years, it is forecasted that the most likely scenario for economic and energy development will follow a moderate growth pattern. Based on the analysis of demand scenarios and projections for a 10-year outlook, the range of electricity consumption volumes to meet the country's consumers is presented in Table 4.19 [18].

At the same time, the combined maximum load of IPS of Ukraine during this period may vary from 23.5 to 30.3 GW.

To ensure synchronous and stable operation of interconnected energy systems, it is crucial to consider the timing and values of the highest and lowest electrical power levels for specific periods in the IPS of Ukraine and neighboring countries that are members of ENTSO-E. The comparison results of these values as of 2018 [15–17] are presented in Table 4.20.

The more detailed parameters of electrical power for February 2, 2018, and June 17, 2018, when maximum and minimum load peaks were observed for each of the energy systems of Ukraine and neighboring countries that are members of ENTSO-E, are organized in Table C.1 [15–17].

Finally, the parameters of participation in covering the total minimum and maximum loads of the ENTSO-E energy system on June 17 and February 28, 2018,

Table 4.19 Minimum and maximum estimate of electricity consumption (gross) for electricity in Ukraine, billion (kWh)

Electricity consumption	2022	2023	2024	2025	2026	2027	2028	2029	2030	2031
– minimum estimate	145	146	148	149	151	152	154	155.5	157	159
– the maximum score	148	150.5	152	154	155	157	159	160.5	162.5	164.5

Table 4.20 The value of the highest and lowest electric power for the IPS of Ukraine and neighboring ENTSO-E member countries in 2018

	Hour of peak load ENTSO-E 28.02.2018 18:00–19:00	Hour of lowest load ENTSO-E 17.06.2018 05:00–06:00
	(MW)	(MW)
Ukraine	23,859	15,112
Poland	24,465	11,842
Romania	8854	4669
Hungary	6542	3389
Slovakia	4413	2496

respectively, for each of the energy systems of Ukraine and neighboring countries that are members of ENTSO-E are summarized in Table C.2.

In Tables 4.19, 4.20, C.1, and C.2, the daily capabilities of electricity import/export from each of the neighboring energy systems are illustrated in a daily perspective (using a typical day as an example), taking into account available capacity and imposed constraints on the available cross-border transmission capacity. From the analysis of these tables, it is evident that each neighboring country exhibits different generation capacities, varying degrees of utilization of their own capacities, and shifts in consumption peaks and troughs. It is worth noting that during periods of maximum daily load in the Ukrainian power system (indicative of potential capacity and reserve deficits), there is a decrease in the maximum potential volumes of import from neighboring countries, due to increased load on the ENTSO-E energy system.

Establishing parallel operation with the powerful European energy system allows Ukraine to reorient its energy system management towards European development vectors, significantly enhancing efficiency and reliability. This transition considers the separation of Ukraine's energy system from the rapidly aging energy system of the Russian Federation. It has enabled Ukraine to facilitate exports and, when necessary, imports of electricity to and from Europe.

The analysis of the current state of generating capacities and operating regimes of the Ukrainian power system has revealed significant challenges in ensuring balance reliability. This is primarily due to the overloading of base-load generating units (NPPs approaching the end of their operational life and the majority of TPPs), acute deficits in maneuvering capacities, and substantial development of renewable energy capacities. Addressing these issues is crucial for preparing Ukraine's energy system for integration with the Continental European ENTSO-E system. It also lays the groundwork for modeling synchronous operation of the Ukrainian power system with neighboring countries' energy systems.

The operation characteristics of the energy systems of neighboring European Union countries—Hungary, Slovakia, Poland, Romania—and Ukraine exhibit varying degrees of utilization of their own capacities and shifts in consumption peaks and troughs. This has allowed for the determination of synchronization conditions, considering the maximum allowable volumes of cross-border electricity exchange, geographic locations, availability and capacity of interconnection lines, and the magnitude of maximum import and export capacities.

4.5 Forecast Balance of Power of IPS of Ukraine for Critical days of Combined Load in 2040

The use of mathematical model [20], constructed for the integrated power system of Ukraine and neighboring countries that are members of ENTSO-E and embedded in the software-information complex [21, 22], allowed forecasting parameters of generating and storage capacities, as well as capacities of cross-border electricity flows. These parameters ensure stable operation of the power systems in synchronous operation modes between Ukraine and neighboring systems (Poland, Romania, and Hungary) within ENTSO-E. Tables 4.21, 4.22, 4.23, 4.24, 4.25 and 4.26 present the calculation results of the system parameters for each of these countries, ensuring stable operation of the integrated energy system.

Table 4.21 Calculation results of the parameters of the **Ukrainian IPS** that will ensure stable operation of the multi-node power system of Ukraine, Poland, and Hungary in synchronous operation mode under the base scenario

Hour	Load	SPP	WPP	NPP	TPP	HPP	PSPP	ESS	CBS
1	18,783	0	292	13,720	2223	2970	– 421	0	0
2	18,012	0	253	13,720	1973	2487	– 421	0	0
3	17,886	0	447	13,720	2152	1987	– 421	0	0
4	17,754	0	739	13,720	1808	1487	0	0	0
5	17,658	0	544	13,720	2082	987	324	0	0
6	18,031	410	642	13,720	2332	603	324	0	0
7	18,253	738	369	13,720	2082	1019	324	0	0
8	18,879	1459	233	13,720	1948	1519	0	0	0
9	20,120	3295	156	13,720	1931	1019	0	0	0
10	20,892	4525	233	13,720	1664	750	0	0	0
11	20,947	5607	311	13,720	1475	250	– 302	0	– 113
12	21,485	5902	603	13,720	1475	250	– 766	0	485
13	21,479	5508	1089	13,720	1475	250	– 766	0	234
14	21,388	5098	933	13,720	1475	250	– 885	155	641
15	21,385	4705	992	13,720	1475	250	– 421	59	605
16	21,212	4885	1031	13,720	1587	750	– 637	0	17
17	21,438	3410	719	13,720	1977	1250	– 216	0	605
18	21,303	2557	467	13,720	2367	1750	– 216	9	650
19	21,737	1197	467	13,720	2757	2250	475	158	650
20	21,479	197	1828	13,720	2873	2750	475	0	– 227
21	22,437	98	1672	13,720	3241	3250	1061	0	– 605
22	22,571	82	817	13,720	2991	3670	586	136	100
23	20,874	0	797	13,720	2601	3170	586	0	0
24	19,511	0	2256	13,720	2211	2670	0	0	– 650
SUM	485,512	49,672	17,889	329,280	50,175	37,588	– 1317	517	2392
MIN	17,658	0	156	13,720	1475	250	– 885	0	– 650
MAX	22,571	5902	2256	13,720	3241	3670	1061	158	650

Table 4.22 Calculation results of the parameters of the **Ukrainian IPS** that will ensure stable operation of the multi-node power system of Ukraine, Poland, and Hungary in synchronous operation mode under the integration scenario

Hour	Load	SPP	WPP	NPP	TPP	HPP	PSPP	ESS	CBS
1	18,783	0	292	14,700	1150	2709	0	0	0
2	18,012	0	253	14,700	1150	2330	– 421	0	0
3	17,886	0	447	14,700	1150	2010	– 421	0	0
4	17,754	0	739	14,700	1200	1510	– 421	26	0
5	17,658	0	544	14,700	1150	1010	324	0	0
6	18,031	410	642	14,700	1150	805	324	0	0
7	18,253	738	369	14,700	1150	971	324	0	0
8	18,879	1459	233	14,700	1175	1268	0	43	0
9	20,120	3295	156	14,700	1150	768	0	52	0
10	20,892	4525	233	14,700	1150	268	0	68	0
11	20,947	5607	311	14,700	1150	250	0	0	– 1003
12	21,485	5902	603	14,700	1150	250	– 507	68	– 680
13	21,479	5508	1089	14,700	1150	250	– 507	0	– 711
14	21,388	5098	933	14,700	1150	250	– 766	0	222
15	21,385	4705	992	14,700	1150	250	– 476	200	– 136
16	21,212	4885	1031	14,700	1150	628	– 476	0	– 355
17	21,438	3410	719	14,700	1343	1128	– 637	19	757
18	21,303	2557	467	14,700	1563	1628	– 421	0	810
19	21,737	1197	467	14,700	1783	2128	– 421	158	1726
20	21,479	197	1828	14,700	1783	2628	475	0	0
21	22,437	98	1672	14,700	1783	3128	1061	0	0
22	22,571	82	817	14,700	1783	3628	1061	136	364
23	20,874	0	797	14,700	1563	3128	586	0	100
24	19,511	0	2256	14,700	1343	2628	0	260	– 1415
SUM	485,512	49,672	17,889	352,800	31,419	35,551	– 1319	1 030	– 321
MIN	17,658	0	156	14,700	1150	250	– 766	0	– 1415
MAX	22,571	5902	2256	14,700	1783	3628	1061	260	1726

Forecasted minimum load day on May 20, 2040, for the power system of Poland. Using the example of forecasted demand values for the minimum load day on May 20, 2040, for the Polish power system, calculations were performed for the base scenario and the 'integration' scenario.

The values of the most significant parameters used in forming the initial data during the mathematical modeling process for the forecasted minimum load day of the Polish power system on May 20, 2040, are provided.

Total generation of the power system for the day: 1,262,390 MWh.

Total load of the power system: MIN—44,880 MW, MAX—59,260 MW.

Table 4.23 Calculation results of the parameters of the **Polish** power system that will ensure stable operation of the multi-node power system of Ukraine, Poland, and Hungary in synchronous operation mode under the base scenario

Hour	Load	SPP	WPP	NPP	TPP	HPP	PSPP	ESS
1	17,125	0	5200	10,597	1230	0	0	0
2	16,306	0	5667	9547	1230	– 236	0	0
3	15,820	0	6206	8522	1230	– 236	0	0
4	15,637	0	6446	8100	1230	– 236	0	0
5	15,061	3	6664	7302	1230	– 236	0	0
6	14,714	268	6052	7303	1230	– 236	0	0
7	15,308	1374	5795	7158	1120	– 236	0	0
8	16,201	3240	3415	8122	920	406	0	0
9	17,509	5435	2003	8848	720	406	0	0
10	18,455	7154	3266	7011	520	406	0	0
11	18,828	8284	5542	4706	320	– 236	0	113
12	18,838	8654	6330	4557	120	– 236	– 100	– 585
13	18,768	8769	6938	4557	120	– 944	0	– 334
14	18,646	8532	8504	4557	120	– 944	0	– 705
15	18,392	7775	8581	4557	120	– 944	0	– 705
16	17,916	6985	8945	4557	120	– 236	0	– 117
17	17,714	5543	8975	5357	230	203	– 58	– 705
18	17,699	4193	8744	6362	430	203	0	– 705
19	18,125	2327	6424	8922	630	203	71	– 550
20	18,568	600	4832	11,362	830	609	– 56	292
21	19,129	75	3818	12,722	1030	609	72	705
22	19,880	0	3539	13,522	1230	812	70	0
23	19,128	0	3903	13,694	1230	203	0	0
24	17,816	0	4211	11,424	1230	203	0	650
SUM	421,584	79,211	140,000	193,366	18,420	– 693	– 1	– 2646
MIN	14,714	0	2003	4557	120	– 944	– 100	– 705
MAX	19,880	8769	8975	13,694	1230	812	72	705

Forecasted power transmission capacities, i.e., permissible cross-border flow capacities of power systems:

- Ukraine—minimum (base scenario) 650 MW, maximum (integration scenario) 2650 MW;
- Poland—minimum 705 MW, maximum 2705 MW;
- Hungary—minimum 100 MW, maximum 2100 MW.

Table 4.24 Calculation results of the parameters of the **Polish** power system that will ensure stable operation of the multi-node power system of Ukraine, Poland, and Hungary in synchronous operation mode under the integration scenario

Hour	Load	SPP	WPP	NPP	TPP	HPP	PSPP	ESS
1	17,125	0	5200	10,597	1230	0	0	0
2	16,306	0	5667	9547	1230	– 236	0	0
3	15,820	0	6206	8522	1230	– 236	0	0
4	15,637	0	6446	8100	1230	– 236	0	0
5	15,061	3	6664	7302	1230	– 236	0	0
6	14,714	268	6052	7303	1230	– 236	0	0
7	15,308	1374	5795	7158	1120	– 236	0	0
8	16,201	3240	3415	8122	920	406	0	0
9	17,509	5435	2003	8848	720	406	0	0
10	18,455	7154	3266	7011	520	406	0	0
11	18,828	8284	5542	4557	320	– 236	0	262
12	18,838	8654	6330	4557	120	– 236	0	– 685
13	18,768	8769	6938	4557	120	– 236	– 89	– 1389
14	18,646	8532	8504	4557	120	– 708	– 100	– 2322
15	18,392	7775	8581	4557	120	– 944	113	– 1908
16	17,916	6985	8945	4557	120	– 944	– 100	– 1745
17	17,714	5543	8975	5357	320	– 236	19	– 2361
18	17,699	4193	8744	6362	520	0	0	– 2218
19	18,125	2327	6424	8802	720	0	8	– 254
20	18,568	600	4832	11,362	920	609	12	133
21	19,129	75	3818	12,722	1120	812	1	483
22	19,880	0	3539	13,522	1230	1015	136	339
23	19,128	0	3903	13,491	1230	406	0	0
24	17,816	0	4211	10,931	1030	203	0	1343
SUM	421,584	79,211	140,000	192,401	18,670	– 693	0	– 10,322
MIN	14,714	0	2003	4557	120	– 944	– 100	– 2361
MAX	19,880	8769	8975	13,522	1230	1015	136	1343

The hourly total power of cross-border flows of the multi-node power system equals zero. Given the specified total load of the power system, cross-border flows will be as follows:

- Ukraine: Export—321 MWh, Import—2392 MWh.
- Poland: Export—2646—10,322 MWh.
- Hungary: Import—255—10,643 MWh.

Table 4.25 Calculation results of the parameters of the **Hungary** power system that will ensure stable operation of the multi-node power system of Ukraine, Poland, and Hungary in synchronous operation mode under the basic scenario

Hour	Load	SPP	NPP	TPP	CBF
1	14,894	0	2352	12,542	0
2	14,270	0	2352	11,918	0
3	13,525	0	2352	11,173	0
4	13,174	0	2352	10,822	0
5	12,948	0	2352	10,596	0
6	12,507	37	2352	10,119	0
7	12,761	338	2352	10,070	0
8	13,736	890	2352	10,494	0
9	14,543	1663	2352	10,528	0
10	15,350	2531	2352	10,467	0
11	15,666	3421	2352	9893	0
12	15,690	3900	2352	9338	100
13	15,737	4268	2352	9017	100
14	15,425	4540	2352	8468	64
15	15,198	4819	2352	7927	100
16	15,070	4378	2352	8240	100
17	15,109	3429	2352	9228	100
18	15,194	2009	2352	10,779	55
19	15,226	692	2352	12,282	– 100
20	15,409	37	2352	13,085	– 64
21	15,639	0	2352	13,387	– 100
22	16,809	0	2352	14,557	– 100
23	16,181	0	2352	13,829	0
24	15,233	0	2352	12,881	0
SUM	355,294	36,951	56,448	261,640	255
MIN	12,507	0	2352	7927	– 100
MAX	16,809	4819	2352	14,557	100

Ukrainian energy system.
Total generation for the day: 485,512 MWh.
Load: MIN—17,658 MW, MAX—22,571 MW.
The optimal number of NPP units, determined in the model with integer variables:

a. Base scenario—14 units, total hourly generation in the base scenario—13,720 MW, total daily generation of NPPs—329,280 MWh;
b. Integration scenario—15 units, total hourly generation in the integration scenario—14,700 MW, total daily generation of NPPs—352 800 MWh.

Table 4.26 Calculation results of the parameters of the **Hungary** power system that will ensure stable operation of the multi-node power system of Ukraine, Poland, and Hungary in synchronous operation mode under the integration scenario

Hour	Load	SPP	NPP	TPP	CBF
1	14,894	0	2352	12,542	0
2	14,270	0	2352	11,918	0
3	13,525	0	2352	11,173	0
4	13,174	0	2352	10,822	0
5	12,948	0	2352	10,596	0
6	12,507	37	2352	10,119	0
7	12,761	338	2352	10,070	0
8	13,736	890	2352	10,494	0
9	14,543	1663	2352	10,528	0
10	15,350	2531	2352	10,467	0
11	15,666	3421	2352	9152	741
12	15,690	3900	2352	8073	1365
13	15,737	4268	2352	7017	2100
14	15,425	4540	2352	6433	2100
15	15,198	4819	2352	5983	2044
16	15,070	4378	2352	6240	2100
17	15,109	3429	2352	7724	1604
18	15,194	2009	2352	9426	1408
19	15,226	692	2352	13,654	− 1472
20	15,409	37	2352	13,154	− 133
21	15,639	0	2352	13,770	− 483
22	16,809	0	2352	15,160	− 703
23	16,181	0	2352	13,929	− 100
24	15,233	0	2352	12,810	72
SUM	355,294	36,951	56,448	251,254	10,643
MIN	12,507	0	2352	5983	− 1472
MAX	16,809	4819	2352	15,160	2 100

Maximum aggregated power of TPPs:

a. Base scenario—3241 MW, total daily generation of TPP—50,175 MWh;
b. Integration scenario—1783 MW, total daily generation of TPP—31,419 MWh.

Maximum aggregated power of hydroelectric power plants (HPP):

a. Base scenario—3670 MW, total daily generation of HPP—37,588 MWh;
b. Integration scenario—3628 MW, total daily generation of HPP—35,551 MWh.

Maximum aggregated generation capacity of pumped-storage power plants (PSPP)—1061 MW.

Installed capacity of WPP—7000 MW. Peak capacity—2356 MW. Total daily generation of WPP—17,889 MWh.

Installed capacity of SPPs—11,000 MW. Peak capacity—5902 MW. Total daily generation of SPP—49,672 MWh.

Installed capacity of energy storage systems—1000 MW with a total capacity of 4000 MWh. Peak generation capacity—base scenario—158 MW, integration scenario—260 MW.

Cross-border flows (CBS):

a. Base scenario—available transfer capacity—650 MW, total daily import—2392 MWh;
b. Integration scenario—available transfer capacity—2650 MW, total daily export—321 MWh.

Power system of Poland.

Total generation for the day: 421,584 MWh.

Load: MIN—14,714 MW, MAX—19,880 MW.

In the forecasted calculation of the structure of Poland's power system for this day, the number of engaged NPP units is 0.

Maximum aggregated power of TPPs:

a. Base scenario—13,694 MW, total daily generation of TPP—193,366 MWh;
b. Integration scenario—13,522 MW, total daily generation of TPP—192,401 MWh.

Maximum aggregated power of hydroelectric power plants (HPP):

a. Base scenario—1230 MW, total daily generation of HPP—18,420 MWh;
b. Integration scenario—1230 MW, total daily generation of HPP—18,670 MWh.

Maximum aggregated generation capacity of pumped-storage power plants (PSPP)—812 MW.

Installed capacity of wind power plants (WPP)—16 00 MW. Peak capacity—8975 MW. Total daily generation of WPP—140,000 MWh;

Installed capacity of SPP—9800 MW. Peak capacity—8769 MW. Total daily generation of SPP—79,211 MWh.

Installed capacity of energy storage systems (ESS)—1000 MW with a total capacity of 4000 MWh. Peak generation capacity—base scenario—72 MW, integration scenario—136 MW.

Cross-border flows (CBF):

a. Base scenario—available transfer capacity—705 MW, total daily export—2646 MWh;
b. Integration scenario—available transfer capacity—2705 MW, total daily export—10,322 MWh.

Energy system of Hungary.

Total generation for the day: 355,294 MWh.

Load: MIN—12,507 MW, MAX—16,809 MW.

The optimal number of engaged NPP units, determined in the model with integer variables:

a. Base scenario—2, total hourly generation in base mode—2352 MW, total daily generation of NPPs—56,448 MWh;
b. Integration scenario—2, total hourly generation in base mode—2352 MW, total daily generation of NPPs—56,448 MWh.

Maximum aggregated power of thermal power plants (TPP):

a. Base scenario—14,557 MW, total daily generation of TPP—261,640 MWh;
b. Integration scenario—15,160 MW, total daily generation of TPP—251,254 MWh.

The forecasted structure of Hungary's power system does not include the use of WPPs.

Installed capacity of SPPs—7590 MW. Peak capacity—4819 MW. Total daily generation of SPP—36,951 MWh.

The forecasted structure of Hungary's power system does not include the use of energy storage systems.

Cross-border flows (CBF):

a. Base scenario—available transfer capacity 100 MW, total import per day—255 MWh;
b. Integration scenario—available transfer capacity—2100 MW, total import per day—10,643 MWh.

The day of the projected maximum load of the Integrated Power System (IPS) of Ukraine on December 20, 2040.

Based on the forecasted demand values for the day of maximum load in 2040 for the IPS of Ukraine, calculations of the parameters of generating and storage capacities, as well as the capacities of cross-border flows, have been performed. These calculations ensure the stable operation of the IPS in synchronous modes with the power systems of Ukraine and neighboring power systems (Poland and Hungary) that are part of ENTSO-E for the "base" and "integration" scenarios.

Tables 4.27, 4.28, 4.29, 4.30, 4.31 and 4.32 present the calculation results of the UES parameters for each country, ensuring the stable operation of the UES.

The values of the most significant parameters used to form the input data in the process of mathematical modeling for the day of the projected maximum load of the IPS of Ukraine on December 20, 2040.

The total generation of the UES per day is 1,943,671 MWh. The total load of the UES ranges from a minimum (MIN) of 65,979 MW to a maximum (MAX) of 89,865 MW.

Interconnectors—cross-border flows.

Table 4.27 Results of calculating the parameters for the **Ukrainian** energy system to ensure the stable operation of the multi-node UES of Ukraine, Poland, and Hungary in synchronous mode under the "base case" scenario

Hour	Load	SPP	WPP	NPP	TPP	HPP	PSPP	ESS	CBF
1	27,364	0	6164	12,740	10,063	250	– 464	0	– 650
2	26,558	0	6164	12,740	10,063	250	– 885	– 500	– 650
3	26,378	0	5406	12,740	10,063	250	– 885	0	– 650
4	26,407	0	4725	12,740	10,063	250	– 421	0	– 650
5	26,763	0	4744	12,740	10,063	250	37	0	– 650
6	27,847	0	4783	12,740	10,313	250	37	71	– 348
7	30,045	148	4706	12,740	12,043	250	37	95	28
8	31,607	279	4511	12,740	13,773	250	– 216	271	0
9	33,204	459	4803	12,740	15,489	250	– 216	– 237	– 84
10	33,934	705	4278	12,740	15,781	505	– 259	146	40
11	34,187	902	3617	12,740	15,862	1005	– 43	0	104
12	33,865	1049	3150	12,740	15,862	1054	– 43	53	0
13	33,647	1033	3033	12,740	15,862	554	324	101	0
14	33,811	754	3461	12,740	15,862	890	685	– 200	– 382
15	33,534	541	2897	12,740	15,862	1390	685	12	– 594
16	34,058	0	3033	12,740	15,862	1890	361	11	0
17	34,317	0	2975	12,740	15,862	2390	– 680	18	650
18	33,849	0	3111	12,740	15,862	2250	– 680	0	566
19	33,206	0	2275	12,740	15,862	1750	– 680	158	357
20	32,751	0	2139	12,740	15,862	1250	512	0	248
21	32,113	0	1731	12,740	15,862	750	663	0	367
22	31,564	0	1906	12,740	15,862	250	663	0	143
23	29,744	0	1692	12,740	14,911	250	151	0	0
24	28,827	0	1536	12,740	14,301	250	0	0	0
SUM	749,582	5869	86,839	305,760	337,270	18,678	– 1 317	– 1	– 2 155
MIN	26,378	0	1536	12,740	10,063	250	– 885	– 500	– 650
MAX	34,317	1049	6164	12,740	15,862	2390	685	271	650

The projected interconnector capacity, i.e., the permissible capacities of cross-border flows between power systems, are as follows:

- Ukraine: minimum 650 MW, maximum 2650 MW;
- Poland: minimum 705 MW, maximum 2705 MW;
- Hungary: minimum 100 MW, maximum 2100 MW.

The hourly total capacity of cross-border flows of the multi-node UES equals zero.

Table 4.28 Results of calculating the parameters for the **Ukrainian** energy system to ensure the stable operation of the multi-node UES of Ukraine, Poland, and Hungary in synchronous mode under the "integration case" scenario

Hour	Load	SPP	WPP	NPP	TPP	HPP	PSPP	ESS	CBF
1	27,364	0	6164	14,700	9085	250	– 302	0	– 2532
2	26,558	0	6164	14,700	9085	250	– 723	– 267	– 2650
3	26,378	0	5406	14,700	9085	250	– 940	0	– 2122
4	26,407	0	4725	14,700	9085	250	– 1058	0	– 1294
5	26,763	0	4744	14,700	9085	250	– 680	– 161	– 1175
6	27,847	0	4783	14,700	9085	250	– 464	0	– 507
7	30,045	148	4706	14,700	10,232	250	– 43	54	0
8	31,607	279	4511	14,700	11,812	250	0	106	– 50
9	33,204	459	4803	14,700	13,079	250	0	– 4	– 84
10	33,934	705	4278	14,700	13,678	250	324	0	0
11	34,187	902	3617	14,700	14,298	250	324	97	0
12	33,865	1049	3150	14,700	14,446	250	324	– 55	0
13	33,647	1033	3033	14,700	14,446	560	0	43	– 169
14	33,811	754	3461	14,700	14,446	896	0	0	– 446
15	33,534	541	2897	14,700	14,446	1396	475	0	– 921
16	34,058	0	3033	14,700	14,446	1896	663	63	– 744
17	34,317	0	2975	14,700	14,446	1713	663	0	– 180
18	33,849	0	3111	14,700	14,446	2002	188	– 34	– 565
19	33,206	0	2275	14,700	14,446	2319	– 421	8	– 120
20	32,751	0	2139	14,700	14,446	1880	– 421	7	0
21	32,113	0	1731	14,700	14,446	1422	– 421	143	92
22	31,564	0	1906	14,700	13,638	922	398	0	0
23	29,744	0	1692	14,700	12,532	422	398	0	0
24	28,827	0	1536	14,700	11,943	250	398	0	0
SUM	749,582	5869	86,839	352,800	300,182	18,678	– 1318	0	– 13,467
MIN	26,378	0	1536	14,700	9085	250	– 1058	– 267	– 2650
MAX	34,317	1049	6164	14,700	14,446	2319	663	143	92

With the specified total load of the UES, the cross-border flows will be:

- Ukraine—Export: 13,467 MWh
- Poland—Import: 3440 MWh
- Hungary—Import: 10,028 MWh.

Ukrainian energy system.

The total generation for the day is 749,582 MWh.
Load MIN—26,378 MW, MAX—34,317 MW.

Table 4.29 Results of calculating the parameters for the **Poland** energy system to ensure the stable operation of the multi-node UES of Ukraine, Poland, and Hungary in synchronous mode under the "base case" scenario

Hour	Load	SPP	WPP	NPP	TPP	HPP	PSPP	ESS	CBF
1	24,098	0	9052	4704	8464	1230	0	0	648
2	22,672	0	9026	4704	7296	1230	0	− 16	431
3	22,019	0	8371	4704	7505	1107	0	− 84	415
4	21,650	0	8329	4704	7366	907	0	− 100	445
5	21,606	0	7995	4704	8231	715	− 472	0	433
6	21,966	0	7691	4704	9175	915	− 708	0	189
7	22,924	0	7053	4704	11,180	1030	− 944	− 100	0
8	26,122	3	7182	4704	13,596	1230	− 472	0	− 122
9	28,550	234	7088	4704	15,531	1230	− 236	0	0
10	29,512	896	7113	4704	15,592	1230	− 236	109	104
11	29,802	1533	6848	4704	15,592	1230	− 236	0	131
12	29,734	1808	6309	4704	15,592	1230	− 236	92	235
13	30,026	1672	6351	4704	15,592	1230	609	− 100	− 33
14	30,076	1079	6715	4704	15,592	1230	609	− 1	147
15	30,386	383	7109	4704	15,592	1230	812	13	543
16	29,882	54	7665	4704	15,592	1230	406	188	0
17	30,084	0	9305	4704	15,244	1230	406	− 100	− 705
18	30,829	0	10,032	4704	15,455	1230	203	− 58	− 705
19	30,620	0	10,434	4704	15,592	1230	− 472	158	− 592
20	30,224	0	10,289	4704	15,245	1230	− 472	0	− 483
21	30,073	0	9857	4704	14,887	1230	− 472	0	− 132
22	29,408	0	8915	4704	14,153	1230	406	0	0
23	27,616	0	8556	4704	12,720	1230	406	0	0
24	25,658	0	8645	4704	10,673	1230	406	0	0
SUM	655,538	7663	195,930	112,896	311,457	28,044	− 693	1	949
MIN	21,606	0	6309	4704	7296	715	− 944	− 100	− 705
MAX	30,829	1808	10434	4704	15,592	1230	812	188	648

The optimal number of engaged NPP units, determined in the model with integer variables:

a. Base Case—13 units, total hourly generation in base mode—12,740 MW, total daily generation by NPPs—305,760 MWh;
b. Integration Case—15 units, total hourly generation in base mode—14,700 MW, total daily generation by NPPs—352,800 MWh.

Table 4.30 Results of calculating the parameters for the **Poland** energy system to ensure the stable operation of the multi-node UES of Ukraine, Poland, and Hungary in synchronous mode under the "integrated case" scenario

Hour	Load	SPP	WPP	NPP	TPP	HPP	PSPP	ESS	CBF
1	24,098	0	9052	4704	8115	1230	0	0	998
2	22,672	0	9026	4704	7000	1230	− 236	− 100	1048
3	22,019	0	8371	4704	6643	1111	− 236	− 7	1432
4	21,650	0	8329	4704	6923	911	− 236	− 68	1088
5	21,606	0	7995	4704	7723	711	− 236	− 100	809
6	21,966	0	7691	4704	9407	911	− 944	0	197
7	22,924	0	7053	4704	11,180	1030	− 944	− 100	0
8	26,122	3	7182	4704	13,740	1230	− 708	42	− 71
9	28,550	234	7088	4704	15,088	1230	0	207	0
10	29,512	896	7113	4704	15,592	1230	0	− 23	0
11	29,802	1533	6848	4704	15,587	1230	0	− 100	0
12	29,734	1808	6309	4704	15,592	1230	0	19	71
13	30,026	1672	6351	4704	15,592	1230	203	182	91
14	30,076	1079	6715	4704	15,592	1230	609	0	146
15	30,386	383	7109	4704	15,592	1230	812	21	534
16	29,882	54	7665	4704	15,592	1230	609	27	0
17	30,084	0	9305	4704	15,592	1230	203	0	− 950
18	30,829	0	10,032	4704	15,592	1230	0	0	− 697
19	30,620	0	10,434	4704	15,592	1230	− 236	0	− 670
20	30,224	0	10,289	4704	15,020	1230	− 236	0	− 494
21	30,073	0	9857	4704	14,611	1230	− 236	0	− 92
22	29,408	0	8915	4704	14,153	1230	406	0	0
23	27,616	0	8556	4704	12,720	1230	406	0	0
24	25,658	0	8645	4704	10,673	1230	406	0	0
SUM	655,538	7663	19,5930	112,896	308,911	28,044	− 594	0	3440
MIN	21,606	0	6309	14,700	6643	711	− 944	− 100	− 950
MAX	30,829	1808	10,434	14,700	15,592	1230	812	207	1432

Maximum aggregated power of TPPs:

a. Base Case—15,862 MW, total daily generation by TPPs—337,270 MWh;
b. Integration Case—14,446 MW, total daily generation by TPPs—300,182 MWh.

Maximum aggregated power of HPPs:

a. Base Case—2390 MW, total daily generation by HPPs—18,678 MWh;
b. Integration Case—2319 MW, total daily generation by HPPs—18,678 MWh.

Table 4.31 Calculation results of the parameters of the **Hungary** power system that will ensure stable operation of the multi-node power system of Ukraine, Poland, and Hungary in synchronous operation mode under the basic scenario

Hour	Load	SPP	NPP	TPP	CBF
1	19,747	0	4704	15,043	0
2	18,794	0	4704	13,990	100
3	17,910	0	4704	13,106	100
4	17,441	0	4704	12,737	0
5	17,452	0	4704	12,748	0
6	18,371	0	4704	13,667	0
7	20,760	0	4704	16,056	0
8	22,952	103	4704	18,145	0
9	23,243	441	4704	18,098	0
10	23,515	949	4704	17,962	− 100
11	23,701	1310	4704	17,787	− 100
12	23,610	1339	4704	17,667	− 100
13	23,893	1118	4704	18,171	− 100
14	23,674	670	4704	18,401	− 100
15	23,338	235	4704	18,498	− 100
16	23,470	15	4704	18,752	− 1
17	23,829	0	4704	18,752	100
18	23,958	0	4704	18,752	100
19	23,500	0	4704	18,752	44
20	23,213	0	4704	18,509	0
21	22,404	0	4704	17,709	− 9
22	21,852	0	4704	17,248	− 100
23	21,051	0	4704	16,347	0
24	20,302	0	4704	15,598	0
SUM	521,980	6181	112,896	402,495	− 266
MIN	17,441	0	4704	12,737	− 100
MAX	23,958	1339	4704	18,752	100

The maximum aggregate generation capacity of the Pumped Storage Power Plant (PSPP) is 685 and 653 MW.

The installed capacity of the WPP is 7000 MW. The peak power is 6164 MW. The total generation of WPP is 86,839 MWh per day;

The installed capacity of the SPP is 11,000 MW. The peak power is 1049 MW. The total generation of SPP is 5869 MWh per day.

Installed power of storage systems—1000 MW with a total capacity of 4000 MWh. Peak generation power:

Table 4.32 Calculation results of the parameters of the **Hungary** power system that will ensure stable operation of the multi-node power system of Ukraine, Poland, and Hungary in synchronous operation mode under the integrated scenario

Hour	Load	SPP	NPP	TPP	CBF
1	19,747	0	4704	14,136	1534
2	18,794	0	4704	13,085	1602
3	17,910	0	4704	13,085	690
4	17,441	0	4704	13,085	206
5	17,452	0	4704	12,937	366
6	18,371	0	4704	13,940	310
7	20,760	0	4704	16,715	0
8	22,952	103	4704	18,752	122
9	23,243	441	4704	18,752	84
10	23,515	949	4704	18,609	0
11	23,701	1310	4704	18,439	0
12	23,610	1339	4704	18,388	− 71
13	23,893	1118	4704	18,752	78
14	23,674	670	4704	18,752	300
15	23,338	235	4704	18,752	387
16	23,470	15	4704	18,752	744
17	23,829	0	4704	18,752	1130
18	23,958	0	4704	18,752	1262
19	23,500	0	4704	18,752	790
20	23,213	0	4704	18,752	494
21	22,404	0	4704	18,411	0
22	21,852	0	4704	17,842	0
23	21,051	0	4704	17,015	0
24	20,302	0	4704	16,243	0
SUM	521,980	6181	112,896	409,450	10,028
MIN	17,441	0	4704	12,937	− 71
MAX	23,958	1339	4704	18,752	1602

- Base Case—271 MW
- Integration Case—328 MW.

Transboundary flows:

- Base Case—available crossing capacity—650 MW, total daily export—2155 MWh;
- Integration Case—available crossing capacity—2650 MW, total daily export—13,467 MWh.

Power system of Poland.

Total daily generation is 655,538 MWh.

The minimum and maximum loads are 21,606 MW and 30,829 MW, respectively.

The optimal parameters for the use of NPPs, as determined by the model with integer variables, are the same for both scenarios: the number of nuclear units is 4, with a total hourly generation of 4740 MW and a total daily generation of 112,896 MWh.

Maximum aggregated power of TPPs:

a. Base Case—15,592 MW, total daily generation by TPPs—311,457 MWh;
b. Integration Case—15,592 MW, total daily generation by TPPs—308,911 MWh.

Maximum aggregated power of HPPs:

a. Base Case—1230 MW, total daily generation by HPPs—28,044 MWh;
b. Integration Case—1230 MW, total daily generation by HPPs—28,044 MWh.

The maximum aggregate generation capacity of the Pumped Storage Power Plant (PSPP) is 812 MW.

The installed capacity of the SPP is 9800 MW. The peak power is 1808 MW. The total generation of SPP is 7633 MWh per day.

Installed power of storage systems—1000 MW with a total capacity of 4000 MWh. Peak generation power:

- Base Case—188 MW
- Integration Case—207 MW.

Transboundary flows:

- Base Case—available crossing capacity—705 MW, total daily import—949 MWh;
- Integration Case—available crossing capacity—2705 MW, total daily import—3440 MWh.

Energy system of Hungary.

Total generation for the day: 521,980 MWh.

Load: MIN—17,441 MW, MAX—23,958 MW.

The optimal number of engaged NPP units, determined in the model with integer variables:

a. Base scenario—4, total hourly generation in base mode—4704 MW, total daily generation of NPPs—112,896 MWh;
b. Integration scenario—4, total hourly generation in base mode—4704 MW, total daily generation of NPPs—112,896 MWh.

Maximum aggregated power of TPP:

a. Base scenario—18,752 MW, total daily generation of TPP—402,495 MWh;
b. Integration scenario—18,752 MW, total daily generation of TPP—409,450 MWh.

The forecasted structure of Hungary's power system does not include the use of WPPs.

Installed capacity of SPPs—7590 MW. Peak capacity—1339 MW. Total daily generation of SPP—6181 MWh.

The forecasted structure of Hungary's power system does not include the use of energy storage systems.

Cross-border flows (CBF):

a. Base scenario—available transfer capacity 100 MW, total export per day—266 MWh;
b. Integration scenario—available transfer capacity—2100 MW, total import per day—10,028 MWh.

The modeling results indicate that by 2040, the trend in the energy systems of the UES, as confirmed by Tables 4.33, 4.34, 4.35, 4.36, 4.37 and 4.38, will be:

1. Increase and Maximum Utilization of Nuclear Power Capacity;
2. Expansion of Cross-Border Transmission Capacity:** Increasing the capacity of interconnecting high-voltage lines will be beneficial for Ukraine and other

Table 4.33 Generation Volumes for the Day of Minimum Load of the IPS of Ukraine, GWh per Day

	Ukraine		Poland		Hungary	
	Min	Max	Min	Max	Min	Max
NPP	352.8	352.8	112.9	112.9	112.9	112.9
TPP	51.8	51.1	172.9	173.1	281.4	281.1
Exp/Imp	0.33	0.42	– 0.19	– 0.57	0.14	0.15

Table 4.34 Generation Volumes for the Day of Maximum Load of the IPS of Ukraine, GWh per Day

	Ukraine		Poland		Hungary	
	Min	Max	Min	Max	Min	Max
NPP	305.8	352.8	112.9	112.9	112.9	112.9
TPP	337.3	300.2	311.5	308.9	414.1	409.5
Exp/Imp	– 2.16	– 13.5	0.95	3.44	1.21	10.03

Table 4.35 Generation volumes for the day of minimum load of the UES of Poland, GWh per Day

	Ukraine		**Poland**		Hungary	
	Min	Max	Min	Max	Min	Max
NPP	329.3	352.8	0	0	56.4	56.4
TPP	50.2	31.4	193.4	192.4	261.6	251.3
Exp/Imp	2.4	– 0.3	– 2.6	– 10.3	0.26	10.6

Table 4.36 Generation Volumes for the Day of Maximum Load of the UES of Poland, GWh per Day

	Ukraine		**Poland**		Hungary	
	Min	Max	Min	Max	Min	Max
NPP	352.8	352.8	112.9	112.9	112.9	112.9
TPP	274.7	275.3	282.2	287.9	415.2	415.5
Exp/Imp	– 0.12	– 0.72	– 0.58	– 6.07	0.7	6.79

Table 4.37 Generation volumes for the day of minimum load of the UES of Hungary, GWh per Day

	Ukraine		Poland		**Hungary**	
	Min	Max	Min	Max	Min	Max
NPP	329.3	352.8	0	0	28.2	56.4
TPP	55.4	36.0	213.2	215.6	240.4	208.0
Exp/Imp	1.84	2.62	– 1.99	– 6.95	0.16	4.33

Table 4.38 Generation volumes for the day of maximum load of the UES of Hungary, GWh per Day

	Ukraine		Poland		**Hungary**	
	Min	Max	Min	Max	Min	Max
NPP	352.8	352.8	112.9	112.9	112.9	112.9
TPP	264.2	274.5	312.7	312.9	387.4	387.4
Exp/Imp	– 4.55	– 14.9	4.41	10.17	0.14	4.74

countries within the UES. This expansion will support the balance and reliability of each country's energy system, particularly in the context of growing nuclear power and RES.

4.6 Forecast Test Calculations for IPS of Ukraine Considering the Necessary Reserves for the Critical Day of May 28, 2040 (FLOOD)

The calculations were performed using the following forecast parameters of the **IPS of Ukraine**:

Installed Capacity:

WPPs—5 GW, SPP—15 GW, Total Capacity and Power of Energy Storage Systems: Capacity—11 GWh, Power—3400 MW.

Daily Generation:

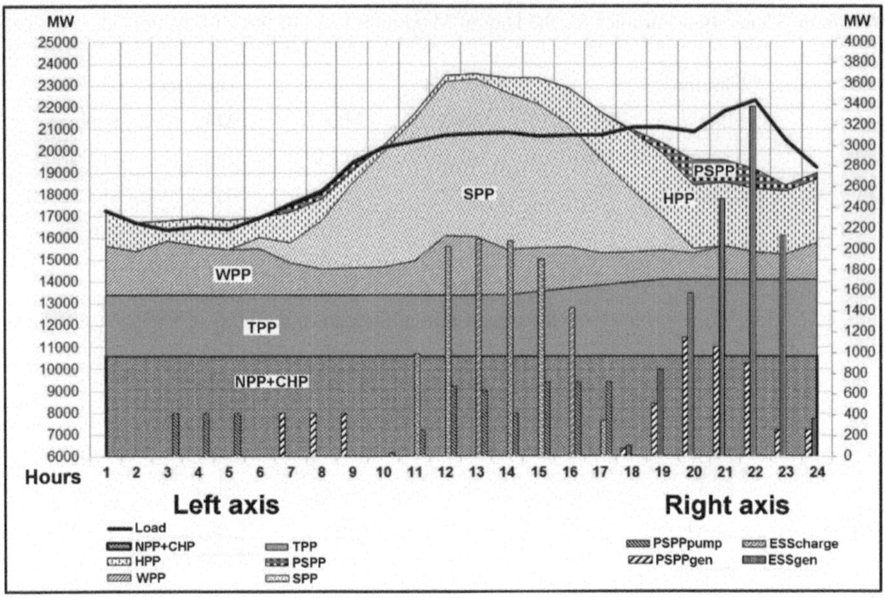

Fig. 4.4 Forecast power balance of the IPS of Ukraine for the Critical Day of May 28, 2040 (FLOOD)

NPP + Combined Heat and Power Plants + Block Stations in constant base mode—10,600 × 24 MWh;

TPPs—73,041 MWh per day (1 × 800, 12 × 300);

Hydroelectric Power Plants (HPP)—37,232 MWh per day;

Maximum Peak Power: WPPs—2720 MW, SPPs—7218 MW.

The results of the forecast calculations are presented in Figs. 4.4, 4.5 and 4.6, and in Tables 4.39, 4.40 and 4.41.

4.7 Forecast Test Calculations for IPS of Ukraine Considering Necessary Reserves for the Critical Day of December 20, 2040 (WINTER)

Forecast test calculations for the critical day of December 20, 2040 (WINTER) have been conducted. These calculations were performed using the following forecast parameters for the IPS of Ukraine:

Installed capacity: WPPs—5 GW, SPPs—15 GW.

Total capacity and power of storage systems: 11 GWh and 2400 MW.

Daily generation: NPP + Combined Heat and Power Plants (CHP) + Block Stations in constant base mode—16,700 × 24 MWh; TPPs—244,193 MWh per day; Hydropower Plants (HPP)—18,679 MWh per day.

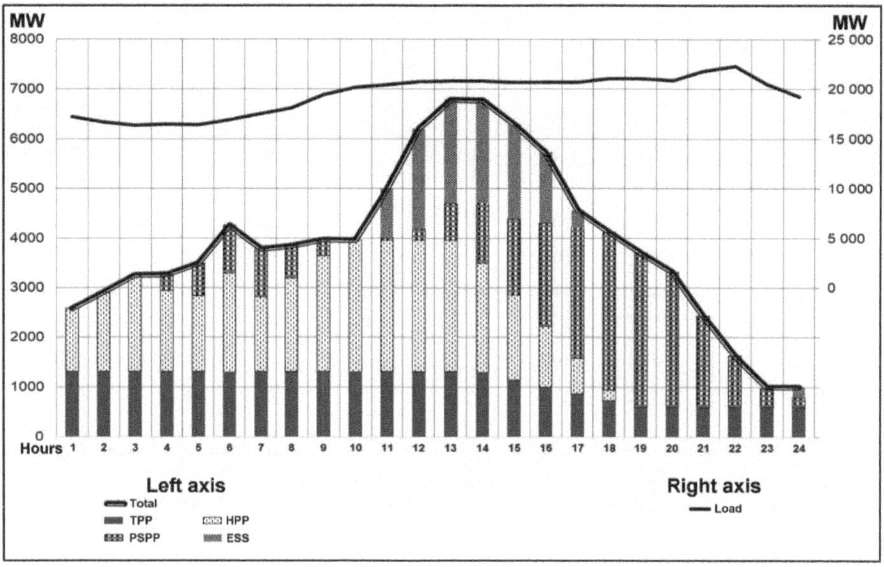

Fig. 4.5 Forecast reserve balance for the load of the IPS of Ukraine for the Critical Day of May 28, 2040 (FLOOD)

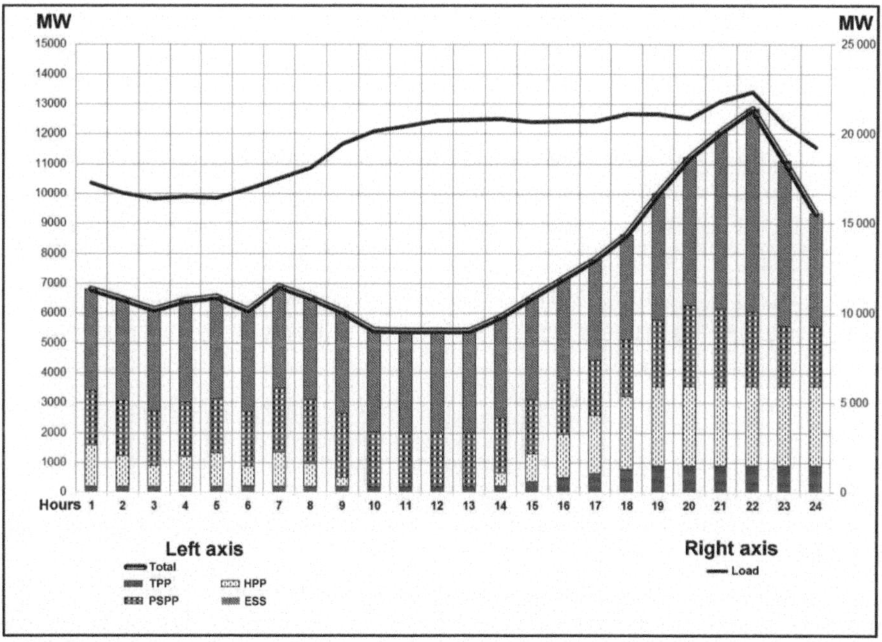

Fig. 4.6 Forecast reserve balance for the unload of the IPS of Ukraine for the Critical Day of May 28, 2040 (FLOOD)

Table 4.39 Results of forecast calculations for the power balance of the IPS of Ukraine for the Critical Day of May 28, 2040 (FLOOD)

Hour	Load	TPP	HPP	PSPP	ESS charge	ESS gen	Reserve for load	Reserve for unload
1	17,277	2800	1637	0	0	0	2573	6812
2	16,706	2800	1293	0	0	0	2917	6468
3	16,395	2800	950	– 421	0	0	3260	6125
4	16,499	2800	1267	– 421	0	0	3267	6442
5	16,416	2801	1370	– 421	0	0	3487	6546
6	16,958	2821	901	0	0	0	4260	6097
7	17,542	2801	1397	421	0	0	3784	6897
8	18,137	2800	1016	421	0	0	3842	6515
9	19,456	2800	561	421	0	0	3973	6060
10	20,171	2808	250	– 43	0	0	3952	5433
11	20,453	2800	250	– 259	988	0	4985	5425
12	20,740	2800	250	– 680	2026	0	6211	5425
13	20,800	2800	250	– 637	2097	0	6794	5425
14	20,838	2830	688	– 421	2077	0	6781	5893
15	20,696	2970	1188	– 723	1907	0	6295	6533
16	20,731	3110	1688	– 723	1430	0	5727	7173
17	20,730	3250	2188	– 723	348	0	4554	7813
18	21,092	3390	2688	86	0	104	4115	8631
19	21,101	3510	2900	507	0	842	3709	10,025
20	20,878	3510	2900	1145	0	1574	3311	11,232
21	21,829	3510	2900	1059	0	2486	2438	12,070
22	22,313	3510	2900	897	0	3374	1639	12,822
23	20,486	3510	2900	260	0	2128	1000	11,101
24	19,270	3510	2900	260	0	365	1000	9338
SUM	467,515	73,041	37,232	5	10,873	10,873	93,874	182,301
MIN	16,395	2800	250		0	0	1000	5425
MAX	22,313	3510	2900		2097	3374	6794	12,822

Maximum peak capacity: WPP—4075 MW, SPP—738 MW.

The results of the forecast calculations are presented in Figs. 4.7, 4.8 and 4.9, and in Tables 4.42, 4.43 and 4.44.

The results of the forecast test calculations, conducted with consideration of the necessary reserves for the critical dates of May 28, 2040 (FLOOD) and December 20, 2040 (WINTER) for the IPS of Ukraine, presented in Tables 4.39, 4.40, 4.41, 4.42, 4.43 and 4.44 and in Figs. 4.4, 4.5, 4.6, 4.7, 4.8 and 4.9, allowed for obtaining numerical estimates of the parameters of power units and technologies necessary for the sustainable operation and prospective development of the power system.

Table 4.40 Forecast reserve balance for the load of the IPS of Ukraine for the Critical Day of May 28, 2040 (FLOOD)

Hour	Load	TPP	HPP	PSPP	ESS	Total reserve for load
1	17,277	1310	1263	0	0	2573
2	16,706	1310	1607	0	0	2917
3	16,395	1310	1950	0	0	3260
4	16,499	1310	1633	324	0	3267
5	16,416	1309	1530	648	0	3487
6	16,958	1289	1999	972	0	4260
7	17,542	1309	1503	972	0	3784
8	18,137	1310	1884	648	0	3842
9	19,456	1310	2339	324	0	3973
10	20,171	1302	2650	0	0	3952
11	20,453	1310	2650	37	988	4985
12	20,740	1310	2650	225	2026	6211
13	20,800	1310	2650	737	2097	6794
14	20,838	1280	2212	1212	2077	6781
15	20,696	1140	1712	1536	1907	6295
16	20,731	1000	1212	2085	1430	5727
17	20,730	860	712	2634	348	4554
18	21,092	720	212	3183	0	4115
19	21,101	600	0	3109	0	3709
20	20,878	600	0	2711	0	3311
21	21,829	600	0	1838	0	2438
22	22,313	600	0	1039	0	1639
23	20,486	600	0	376	24	1000
24	19,270	600	0	188	212	1000
SUM	467,515	25,599	32,368	24,798	11,109	93,874
MIN	16,395	600	0	0	0	1000
MAX	22,313	1310	2650	3183	2097	6794

Thus, numerical studies using the created and improved models made it possible to determine the optimal modes of using generating, reserve, and storage capacities to cover the forecasted electric load schedule at different values of the total installed and peak capacities of SPPs and WPPs.

Table 4.41 Forecast reserve balance for the unload of the IPS of Ukraine for the Critical Day of May 28, 2040 (FLOOD)

Hour	Load	TPP	HPP	PSPP	ESS	Total reserve for unload
1	17,277	200	1387	1825	3400	6812
2	16,706	200	1043	1825	3400	6468
3	16,395	200	700	1825	3400	6125
4	16,499	200	1017	1825	3400	6442
5	16,416	201	1120	1825	3400	6546
6	16,958	221	651	1825	3400	6097
7	17,542	201	1147	2149	3400	6897
8	18,137	200	766	2149	3400	6515
9	19,456	200	311	2149	3400	6060
10	20,171	208	0	1825	3400	5433
11	20,453	200	0	1825	3400	5425
12	20,740	200	0	1825	3400	5425
13	20,800	200	0	1825	3400	5425
14	20,838	230	438	1825	3400	5893
15	20,696	370	938	1825	3400	6533
16	20,731	510	1438	1825	3400	7173
17	20,730	650	1938	1825	3400	7813
18	21,092	790	2438	1899	3504	8631
19	21,101	910	2650	2223	4242	10,025
20	20,878	910	2650	2698	4974	11,232
21	21,829	910	2650	2624	5886	12,070
22	22,313	910	2650	2488	6774	12,822
23	20,486	910	2650	2013	5528	11,101
24	19,270	910	2650	2013	3765	9338
SUM	467,515	10,641	31,232	47,955	92,473	182,301
MIN	16,395	200	0	1825	3400	5425
MAX	22,313	910	2650	2698	6774	12,822

4.8 Research Using a Mathematical Model for Minimizing the Weighted Average Cost

In the course of research using the mathematical model for minimizing the weighted average cost, model calculations were performed using the actual load graphs of the IPS of Ukraine for 2018 [15, 23]. The calculations took into account ratios (2.45), (2.46), (2.47), and the following technological constraints:

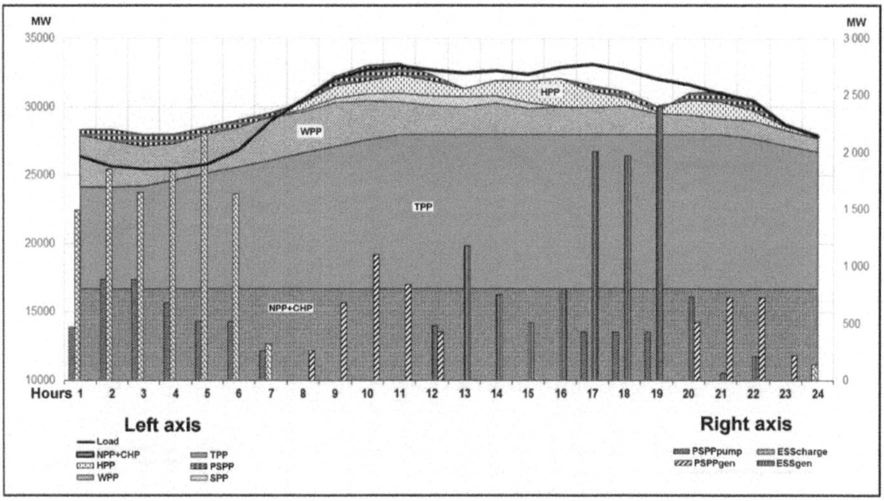

Fig. 4.7 Forecast power balance of IPS of Ukraine for the Critical Day of December 20, 2040 (WINTER)

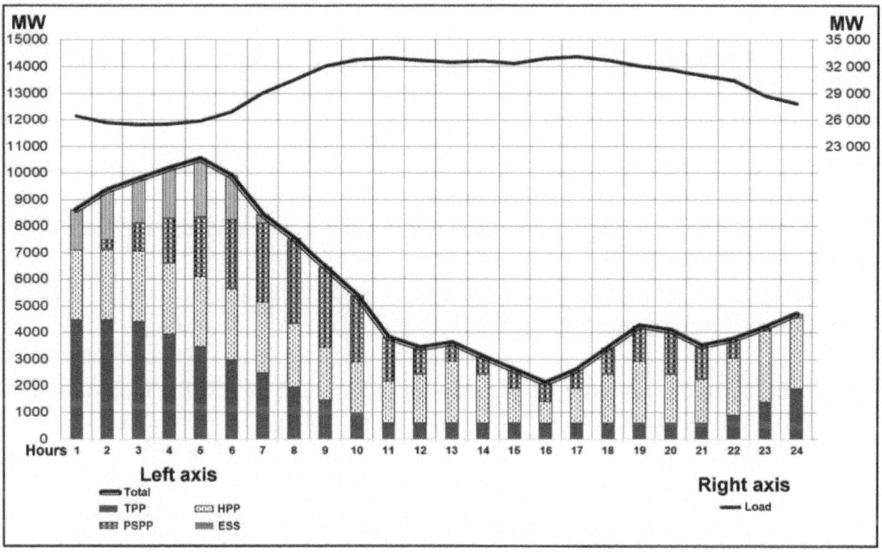

Fig. 4.8 Forecast reserve balance for load of IPS of Ukraine for the Critical Day of December 20, 2040 (WINTER)

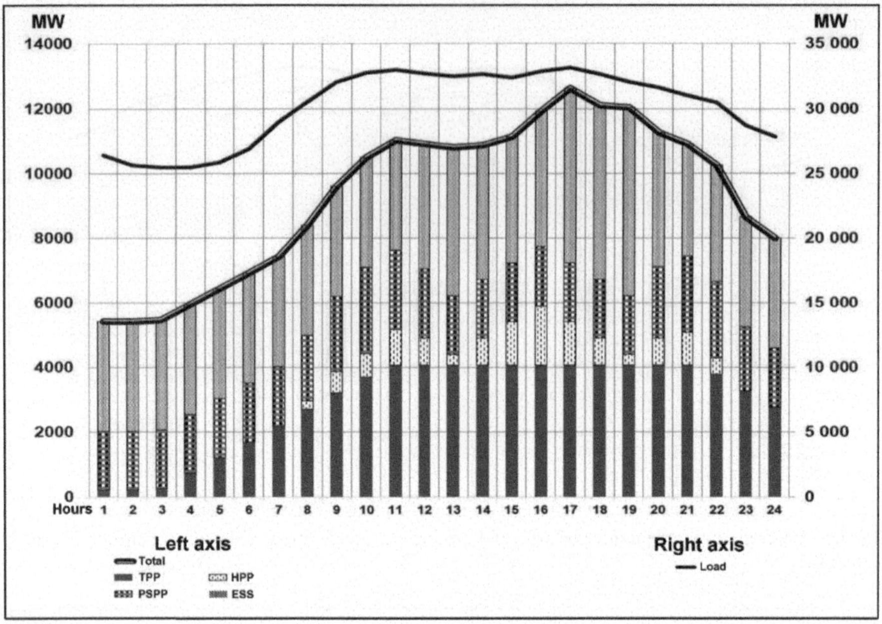

Fig. 4.9 Forecast reserve balance for unloading of IPS of Ukraine for the Critical Day of December 20, 2040 (WINTER)

- Constraints on the total generating capacity of NPPs $P_t^{NPP\_O}$ ensuring the use of NPPs P_{BGM}^{NPP} in the base generation mode: $P_t^{NPP\_O} = P_{BGM}^{NPP}$;
- Constraints on the minimum and maximum total generating capacity of TPPs: $P_{\min}^{TPP} \leq P_t^{TPP\_O} \leq P_{\max}^{TPP}$;
- Constraints on the minimum and maximum total generating capacity of HPPs and pumped storage power plants (PSPPs): $P_{\min}^{HPP} \leq P_t^{HPP\_O} \leq P_{\max}^{HPP}$, $P_{\min}^{HPSS\_G} \leq P_t^{HPSS\_G\_O} \leq P_{\max}^{HPSS\_G}$;
- Constraints on the volumes of generation and storage of PSPPs: the volume of consumption equals the volume of storage, taking into account the loss coefficient α:

$$\sum_t^T \left(P_t^{HPSS\_G\_O} - (1-\alpha)P_t^{HPSS\_P\_O}\right) = \emptyset;$$

- Constraints on import and export volumes: $Z_t^{I\_O} \leq E^Z$; $Z_t^{E\_O} \leq E^Z$; $\forall t \in T$.

These constraints ensure the balance and stability of the power system while optimizing the costs associated with power generation and storage. The use of these models and constraints allows for an effective analysis of the energy system, facilitating strategic decision-making for future developments and optimizations.

Table 4.42 Results of forecast calculations of power balance for IPS of Ukraine for the Critical Day of December 20, 2040 (WINTER)

Hour	Load	TPP	HPP	PSPP	ESS charge	ESS gen	Reserve for load	Reserve for unload
1	26,421	7426	250	– 464	1491	0	8615	5425
2	25,642	7426	250	– 885	1849	0	9334	5425
3	25,469	7480	250	– 885	1649	0	9765	5479
4	25,497	7960	250	– 680	1856	0	10,177	5959
5	25,840	8440	250	– 519	2167	0	10,520	6439
6	26,887	8930	250	– 519	1636	0	9875	6929
7	29,009	9420	250	– 259	315	0	8440	7419
8	30,517	9930	521	260	0	0	7532	8388
9	32,059	10,440	909	681	0	0	6446	9610
10	32,764	10,924	981	1102	0	0	5378	10,490
11	33,008	11,300	1334	842	0	0	3813	11,031
12	32,697	11,300	1076	421	0	476	3423	10,925
13	32,487	11,300	576	0	0	1179	3599	10,804
14	32,645	11,300	1076	0	0	751	3099	10,876
15	32,378	11,300	1576	0	0	506	2599	11,131
16	32,883	11,300	2076	0	0	802	2099	11,927
17	33,134	11,300	1576	– 421	0	2012	2599	12,637
18	32,682	11,300	1076	– 421	0	1970	3423	12,095
19	32,061	11,300	576	– 421	0	2403	4247	12,028
20	31,622	11,300	1076	507	0	734	4071	11,257
21	31,006	11,300	1250	724	0	63	3499	10,911
22	30,475	11,009	750	724	0	208	3741	10,265
23	28,718	10,499	250	217	0	0	4202	8649
24	27,833	10,009	250	0	141	0	4682	8008
SUM	723,734	244,193					135,178	224,107
MIN	25,469	7426					2099	5425
MAX	33,134	11,300					10,520	12,637

The actual power distribution for covering the load graph of the IPS of Ukraine over a modeling horizon of $T = 100$ h (approximately 4 characteristic days) in 2018 is shown in Fig. 4.10. The optimal power distribution for covering the same load graph of the IPS of Ukraine over the modeling horizon T = 100 h is shown in Fig. 4.11.

Based on the optimization calculations, the absolute difference Δ_T^{Σ} and the relative difference Ψ_T^{Σ} between the actual and optimized weighted average cost, calculated using the formulas $\Delta_T^{\Sigma} = C^{\Sigma} - C^{\Sigma\text{-}O}$ and $\Psi_T^{\Sigma} = \frac{\Delta_T^{\Sigma}}{C^{\Sigma}}$, are as follows for the modeling horizon $T = 100$ h (approximately 4 characteristic days) in 2018:

Table 4.43 Forecast reserve balance for load of IPS of Ukraine for the Critical Day of December 20, 2040 (WINTER)

Hour	Load	TPP	HPP	PSPP	ESS	Total reserve for load
1	26,421	4474	2650	0	1491	8615
2	25,642	4474	2650	361	1849	9334
3	25,469	4420	2650	1046	1649	9765
4	25,497	3940	2650	1731	1856	10,177
5	25,840	3460	2650	2243	2167	10,520
6	26,887	2970	2650	2619	1636	9875
7	29,009	2480	2650	2995	315	8440
8	30,517	1970	2379	3183	0	7532
9	32,059	1460	1991	2995	0	6446
10	32,764	976	1919	2483	0	5378
11	33,008	600	1566	1647	0	3813
12	32,697	600	1824	999	0	3423
13	32,487	600	2324	675	0	3599
14	32,645	600	1824	675	0	3099
15	32,378	600	1324	675	0	2599
16	32,883	600	824	675	0	2099
17	33,134	600	1324	675	0	2599
18	32,682	600	1824	999	0	3423
19	32,061	600	2324	1323	0	4247
20	31,622	600	1824	1647	0	4071
21	31,006	600	1650	1249	0	3499
22	30,475	891	2150	700	0	3741
23	28,718	1401	2650	151	0	4202
24	27,833	1891	2650	0	141	4682
SUM	723,734	41,407	50,921	31,746	11,104	135,178
MIN	25,469	600	824	0	0	2099
MAX	33,134	4474	2650	3183	2167	10,520

$$\Delta_T^{\Sigma} = 120\,187\,378\,\text{UAH and } \Psi_T^{\Sigma} = 38.97\%.$$

For a one-year modeling horizon (8760 h), the relative difference $\Psi_{T=1year}^{\Sigma}$ in the weighted average cost ranges from a minimum value of 1.43% to a maximum value of 66.8%, with an average value of 25.44%.

The hourly distribution of the absolute differences δ_t and the relative differences ψ_t between the actual and optimized weighted average cost is presented in

Table 4.44 Forecast Reserve Balance for Unloading of IPS of Ukraine for the Critical Day of December 20, 2040 (WINTER)

Hour	Load	TPP	HPP	PSPP	ESS	Total reserve for unload
1	26,421	200	0	1825	3400	5425
2	25,642	200	0	1825	3400	5425
3	25,469	254	0	1825	3400	5479
4	25,497	734	0	1825	3400	5959
5	25,840	1214	0	1825	3400	6439
6	26,887	1704	0	1825	3400	6929
7	29,009	2194	0	1825	3400	7419
8	30,517	2704	271	2013	3400	8388
9	32,059	3214	659	2337	3400	9610
10	32,764	3698	731	2661	3400	10,490
11	33,008	4074	1084	2473	3400	11,031
12	32,697	4074	826	2149	3876	10,925
13	32,487	4074	326	1825	4579	10,804
14	32,645	4074	826	1825	4151	10,876
15	32,378	4074	1326	1825	3906	11,131
16	32,883	4074	1826	1825	4202	11,927
17	33,134	4074	1326	1825	5412	12,637
18	32,682	4074	826	1825	5370	12,095
19	32,061	4074	326	1825	5803	12,028
20	31,622	4074	826	2223	4134	11,257
21	31,006	4074	1000	2374	3463	10,911
22	30,475	3783	500	2374	3608	10,265
23	28,718	3273	0	1976	3400	8649
24	27,833	2783	0	1825	3400	8008
SUM	723,734	70,769	12,679	47,955	92,704	224,107
MIN	25,469	200	0	1825	3400	5425
MAX	33,134	4074	1826	2661	5803	12,637

Fig. 4.12. These distributions provide insights into the periods where the cost optimization has the most significant impact, highlighting potential areas for operational improvements and cost savings in the power system.

Results of Calculations and Model Validation

The calculations have confirmed the adequacy of the model and its capability for calculating the optimization potential of generating resources in the IPS of Ukraine. Using the model, numerical estimates of the potential for optimizing the operational modes of the IPS components in Ukraine have been obtained. The optimized mode lowers the production cost of electricity. By leveling out the operational graphs,

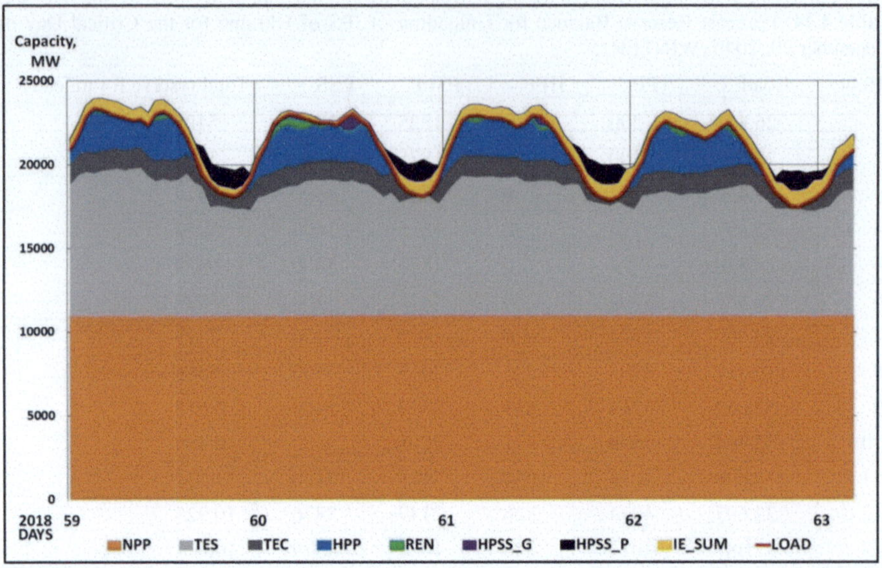

Fig. 4.10 The actual power distribution for covering the load graph of the IPS of Ukraine over a modeling horizon of T = 100 h (approximately 4 characteristic days) in 2018

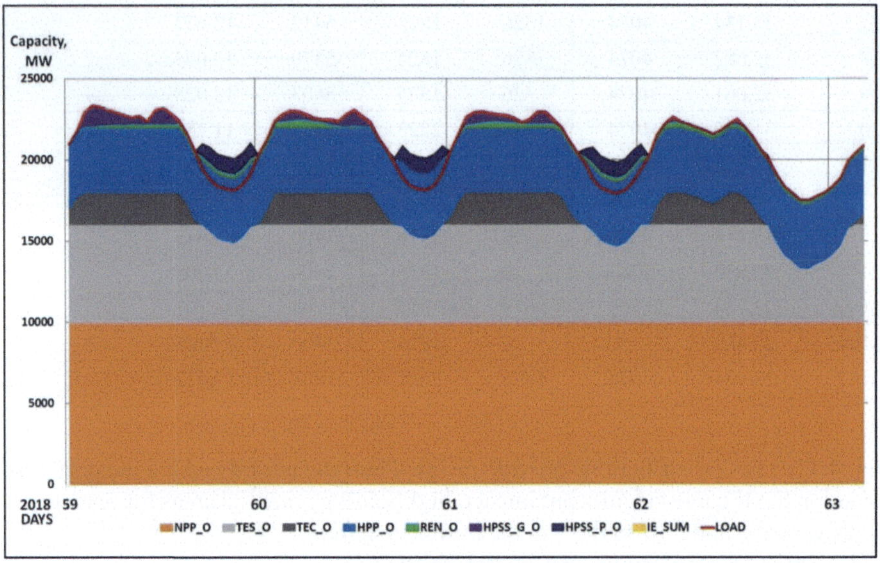

Fig. 4.11 The optimal power distribution for covering the same load graph of the IPS of Ukraine over the modeling horizon T = 100 h (approximately 4 characteristic days) in 2018

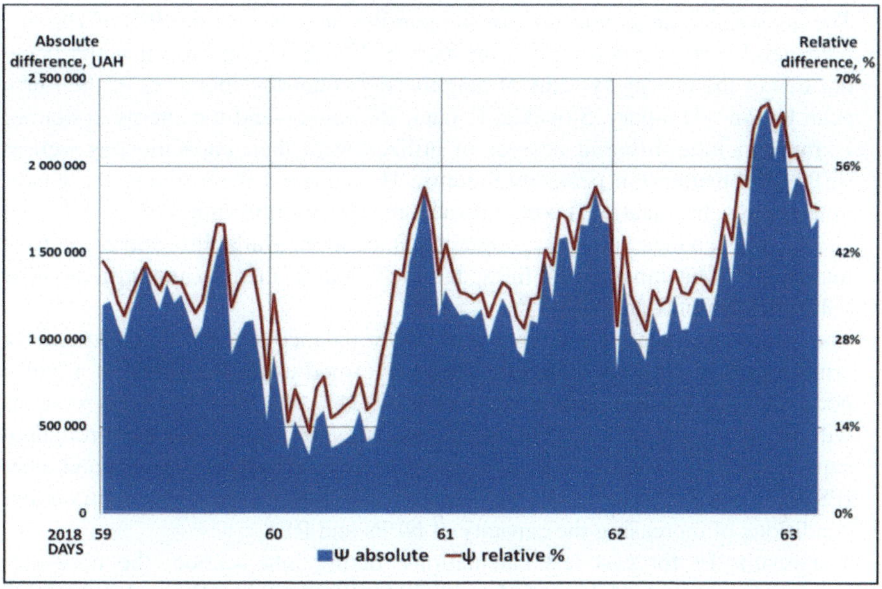

Fig. 4.12 The hourly distribution of the absolute differences δ_t and the relative differences ψ_t between the actual and optimized weighted average cost for the modeling horizon $T = 100$ h (approximately 4 characteristic days) in 2018

the demands on maneuvering modes are lessened. The operational and technological characteristics are enhanced, reducing the probability of emergency repairs and extending the time between scheduled maintenance.

This model can serve as a foundation for further development of a family of optimization models for the structure of generating and storage capacities. Additionally, it can contribute to the modernization and development of software and information systems for modeling the IPS of Ukraine.

4.9 Conclusions

1. Using the new dynamic model of cyclical development of generation systems in energy, scenarios for the development of SPPs in Germany and Ukraine were calculated.
2. The method of economic and technological forecasting of optimal functioning and development parameters of the energy system, and the model of long-term technological renewal of the energy system structure allowed making forecasts of the structure of consumption and generation of electricity for the IPS of Ukraine until 2040, enabling an assessment of the potential for improving the components of the IPS of Ukraine.

3. The analysis of the conditions for the parallel operation of the IPS of Ukraine with neighboring countries that are part of ENTSO-E showed that the peculiarities of the energy systems of neighboring countries, members of the European Union—Hungary, Slovakia, Poland, Romania—and the energy system of Ukraine include different degrees of utilization of their capacities, as well as shifts in consumption peaks/minimums. This allowed determining the conditions for synchronization, taking into account the limiting values of cross-border electricity exchange volumes, geographical location, availability and capacity of interstate power transmission lines, and the values of the maximum permissible volumes of import and export.
4. The results of modeling the forecasted power balance of the IPS of Ukraine for critical days of combined load in 2040, performed using the model of a multi-node integrated power system, show that by 2040, the trend in energy systems will be: increasing and maximizing the use of NPPs; increasing the throughput capacity of interstate power lines, which is beneficial for both Ukraine and other IPS countries, as it will ensure the balance reliability of each UES country under conditions of increasing the capacity of NPPs and RES.
5. The results of forecast test calculations, taking into account the necessary reserves for the critical days of May 28, 2040 (FLOOD) and December 20, 2040 (WINTER) for the IPS of Ukraine, provided numerical estimates of the parameters of power units and technologies required for the sustainable operation and prospective development of the power system.
6. The modeling results indicate that despite technical progress in renewable energy, particularly the reduction in the cost of electricity production by wind and photovoltaic power plants, due to the variability of their electricity generation, they can only be integrated into energy associations if there are sufficiently powerful, energy-efficient, and economically effective industrial energy storage systems. Also, promising areas for their application are local energy consumption systems that can use electricity with parameters that do not meet the stability requirements of voltage and frequency imposed on energy associations that are components of the IPS of Ukraine.
7. The test calculations showed sufficient consistency with known results, providing grounds to consider the proposed new and improved models adequate, and the further development of the models promising for achieving the necessary levels of detail in forecast scenarios for the development and renewal of the components of the IPS of Ukraine in the long-term perspective.

References

1. Marín-Cano, C., Mejía-Giraldo, D.: Levelized avoided cost of electricity model based on power system operation. DYNA **85**(206), 79–84. Home Page. https://doi.org/10.15446/dyna.v85n206.69577. Date of Access: 27 Oct 2023

2. Breyer Christian, C.: PV manufacturing in Europe—European technology and innovation platform photovoltaics. Share and Discover Knowledge on SlideShare. https://www.slideshare.net/cluster_tweed/pv-manufacturing-in-europe-european-technology-and-innovation-platform-photovoltaics. Date of Access: 09 Aug 2023
3. On the Ratification of the Association Agreement between Ukraine, on the one hand, and the European Union, the European Atomic Energy Community, and their Member States, on the other hand: Law of Ukraine No. 1678-VII, dated September 16, 2014. https://zakon.rada.gov.ua/laws/show/1678-18#Text. Accessed 27 July 2023
4. On the Electricity Market: Law of Ukraine No. 2019-VIII, dated April 13, 2017. Updated as of July 27, 2023. https://zakon.rada.gov.ua/laws/show/2019-19#Text. Accessed 27 July 2023
5. Market Rules: Resolution No. 307 dated 14 Mar 2018
6. On Approval of the Market Rules for the Day-Ahead and Intraday Markets: Resolution No. 308 of the National Commission for State Regulation of Energy and Public Utilities dated March 14, 2018, as of December 2, 2022. https://zakon.rada.gov.ua/laws/show/v0308874-18#Text. Accessed 27 July 2023
7. Denysov, V., Kulyk, M., Babak, V., Zaporozhets, A., Kostenko, G.: Modeling Nuclear-Centric Scenarios for Ukraine's Low-Carbon Energy Transition Using Diffusion and Regression Techniques. **Energies**. 17(20), 5229 (2024). https://doi.org/10.3390/en17205229
8. Projected costs of generating electricity 2020—analysis. IEA. https://www.iea.org/reports/projected-costs-of-generating-electricity-2020. Date of Access: 21 July 2023
9. Regulation on Approval of the Transmission System Code. Decision of the National Commission for State Regulation of Energy and Public Utilities No. 309 of March 14, 2018. Updated as of April 14, 2023. Available at: https://zakon.rada.gov.ua/laws/show/v0309874-18#Text. Accessed: 27 July 2023
10. Denisov, V.A., Chuprina, L.V.: Technological conditions synchronous operation of united power system of Ukraine and neighboring ENTSO-E-member power systems. Probl. Gen. Energy **2021**(3), 53–62 (2021). https://doi.org/10.15407/pge2021.03.053. Date of Access: 27. July 2023
11. Hryshko, V.I.: Synchronization of the United Energy System of Ukraine with ENTSO-E: Economic Dimension. Scientific Notes of the University KROK, pp. 9–16 (2022). https://doi.org/10.31732/2663-2209-2022-66-9-16. Accessed: 21 July 2023
12. Presentation by NPC Ukrenergo: Integration Plan of the Ukrainian Power System into ENTSO-E. Share and Discover Knowledge on SlideShare. https://www.slideshare.net/NPCUkrenergo/entsoe-233692950. Accessed 27 July 2023
13. First Victory on the Energy Front: The Energy System of Ukraine Became Part of the European Energy System. Official Portal of the Verkhovna Rada of Ukraine. https://www.rada.gov.ua/print/220582.html. Accessed: 21 July 2023
14. Ministry of Energy of Ukraine. Energy Strategy of Ukraine for the Period up to 2050. https://mev.gov.ua/reforma/enerhetychna-stratehiya. Accessed: 27 July 2023
15. Hourly power balance of Ukraine's IPS for 2018. https://data.gov.ua/dataset
16. Central collection and publication of electricity generation, transportation and consumption data and information for pan-European market. ENTSO-E Transparency Platform. https://transparency.entsoe.eu/load-domain/r2/yearLoad/show. Date of Access: 21 July 2023
17. Balance of Flows Between the Energy System of Ukraine and the Energy Systems of Neighboring Countries. Official Website of NPC "Ukrenergo". https://ua.energy/peredacha-i-dyspetcheryzatsiya/dyspetcherska-informatsiya/peretoky/
18. Report on the Assessment of the Adequacy of Generating Capacities to Meet Forecasted Electricity Demand and Ensure Necessary Reserves for 2022, approved by the NKREKP Resolution No. 664 dated April 11, 2023. NPC "Ukrenergo". Available at: https://ua.energy/zvit-z-otsinky-vidpovidnosti-dostatnosti-generuyuchyh-potuzhnostej/. Accessed: 21 July 2023
19. Cabinet of Ministers of Ukraine.: Approval of the Energy Strategy of Ukraine for the Period up to 2035 "Security, Energy Efficiency, Competitiveness": Order No. 605-p, August 18, 2017 (2017). Available at: https://zakon.rada.gov.ua/laws/show/605-2017-p#Text. Accessed: 7 Aug 2023

20. Denisov, V.: Integrated Power System multi-node model, taking into account the nondispatchable of renewable energy sources. In: 2022 IEEE 8th International Conference on Energy Smart Systems (ESS), Kyiv, Ukraine, 12–14 October 2022 (2022). https://doi.org/10.1109/ess57819.2022.9969255. Date of Access: 27 July 2023
21. Kostenko, G., Zaporozhets, A.: Transition from Electric Vehicles to Energy Storage: Review on Targeted Lithium-Ion Battery Diagnostics. **Energies**. 17(20), 5132 (2024). https://doi.org/10.3390/en17205132
22. Denysov, V.: Software and information complex for district heat supply systems modeling. Syst. Res. Energy **2022**(1), 38–45 2022. https://doi.org/10.15407/srenergy2022.01.038. Date of Access: 27 July 2023
23. Hourly electricity balance of the IPS of Ukraine—Energy Map. Energy Map. https://map.ua-energy.org/en/resources/8998f2ed-379f-4fa9-9076-88782b32ee4f/. Date of Access: 21 July 2023

Appendix A
Key Indicators of Energy Modeling Systems

See Tables A.1, A.2, A.3 and A.4.

A. Zaporozhets et al., *Structure Optimization of Power Systems with Renewable Energy Sources*, Studies in Systems, Decision and Control 583,
https://doi.org/10.1007/978-3-031-83697-8

Table A.1 Key indicators of energy modeling systems

Name of the modeling system	Year of release	Purpose	Time horizon, years	Number of intervals per year
AEOLIUS	2005	M	≤ 2	≥ 8760
AIM	1996	I	> 30	≤ 1
AURORAxmp	2017	Y + I	> 30	≥ 8760
BALMOREL	2000	Y + I	> 1 … ≤ 30	≥ 8760
BCHP Screening tool	2003	Y	≤ 2	≥ 8760
Calliope	2015	Y + I	> 1 … ≤ 30	≥ 8760
CASPOC	2018	M	≤ 2	≥ 8760
CYME	2018	M	> 1 … ≤ 30	≥ 8760
CGEN	2007	Y	≤ 2	≥ 8760
CHP capacity optimizer	2005	Y + I	> 2 … ≤ 30	≥ 8760
CHPSizer	2001	Y	> 30	> 1 … < 8760
COMPETES	2004	Y	≤ 2	> 1 … < 8760
COMPOSE	2008	Y + I	> 30	≥ 8760
COSMEE	1998	Y	≤ 2	≥ 8760
DEARS	2005	I	> 30	≤ 1
DER-CAM	2000	Y + I	≤ 2	≥ 8760
DESSTinEE	2015	M	> 1 … ≤ 30	≥ 8760
DIETER	2015	Y + I	≤ 2	> 1 … < 8760
DIgSILENT/Power factory	2019	M	> 1 … ≤ 30	≥ 8760
DIME	2007	Y + I	> 2 … ≤ 30	> 1 … < 8760
DNE21	1999	I	> 30	≤ 1
DTI	1999	I	> 2 … ≤ 30	> 1 … < 8760
E2M2s	2002	Y + I	> 2 … ≤ 30	> 1 … < 8760
E3database	2008	M	> 2 … ≤ 30	> 1 … < 8760
E4cast	2000	I	> 2 … ≤ 30	≤ 1
EGEAS	1979	Y + I	> 2 … ≤ 30	≥ 8760
Elfin	1991	Y	≤ 2	≥ 8760
EMCAS	2002	I	> 30	≥ 8760
EMELIE	2003	M	> 30	> 1 … < 8760
EMINENT	2005	M	≤ 2	≥ 8760
EMMA	2017	Y + I	> 2 … ≤ 30	≥ 8760
EMPIRE	2016	I	> 30	≤ 1
EMPS	1975	Y	> 2 … ≤ 30	> 1 … < 8760
Energy 2020	1972	M	> 2 … ≤ 30	≤ 1
EnergyPLAN	1999	Y + I	≤ 2	≥ 8760

(continued)

Table A.1 (continued)

Name of the modeling system	Year of release	Purpose	Time horizon, years	Number of intervals per year
energyPRO	1987	Y + I	> 30	≥ 8760
Enertile	2019	Y + I	> 30	≥ 8760
ENTIGRIS	2019	Y + I	> 30	≥ 8760
ERIS	1996	I	> 30	≤ 1
ESPAUT	2005	I	≤ 2	≥ 8760
ETM Model	2018	M	> 30	≥ 8760
Energy Transition Model	2017	M	> 30	≥ 8760
ETSAP-TIAM	2008	Y + I	> 30	> 1 … < 8760
EUCAD	2015	Y	≤ 2	≥ 8760
EUPower-Dispatch	2013	Y	≤ 2	≥ 8760
Ficus	2018	Y + I	≤ 2	≥ 8760
FOSSIL2	1972	I	> 30	> 1 … < 8760
GCAM	2019	M	> 30	≤ 1
GEM-E3	2013	M	> 30	≤ 1
genEris	2005	I	> 30	≤ 1
GENESYS	2014	I	> 30	≥ 8760
GridLAB-D	2019	M	> 2 … ≤ 30	> 1 … < 8760
GTMax	1995	Y	≤ 2	≥ 8760
H2A analysis	2003	Y + I	≤ 2	≥ 8760
H2RES	2000	Y	> 30	≥ 8760
HOMER	1992	Y + I	≤ 2	≥ 8760
HUD CHP Screening Tool	2002	M	≤ 2	> 1 … < 8760
HYDROGEMS	1995	M	≤ 2	≥ 8760
HYPERSIM	2018	M	≤ 2	≥ 8760
HYPRO	2005	M	> 2 … ≤ 30	≤ 1
iHOGA	2019	Y + I	≤ 2	≥ 8760
IKARUS	1991	I	> 30	≤ 1
IMAGE-TIMER	1997	Y + I	> 30	> 1 … < 8760
IMAKUS	2012	Y + I	≤ 2	≥ 8760
INFORSE	2002	M	> 30	≤ 1
Invert	2003	I	> 2 … ≤ 30	≤ 1
INVERT/EE-Lab	2015	M	> 30	≤ 1
IPM	1980	Y + I	> 2 … ≤ 30	≥ 8760
IPSA 2	2015	M	≤ 2	≥ 8760
IRiE	2012	Y	≤ 2	≥ 8760

(continued)

Table A.1 (continued)

Name of the modeling system	Year of release	Purpose	Time horizon, years	Number of intervals per year
LEAP	1980	M	> 30	≤ 1
LIBEMOD	2009	M	> 2 … ≤ 30	> 1 … < 8760
LIMES-EU	2011	M	> 30	≤ 1
LOADMATCH	2015	M	> 2 … ≤ 30	≥ 8760
LUSYM	2014	Y	≤ 2	≥ 8760
MARKAL/TIMES	1978	Y + I	> 30	> 1 … < 8760
MELP	1997	I	> 2 … ≤ 30	≤ 1
MERGE	1994	M	> 30	≤ 1
Mesap PlaNet	1997	M	> 30	≥ 8760
MESSAGE	1978	Y + I	> 30	≤ 1
MiniCAM	1982	M	> 30	≤ 1
MTSIM	2009	Y	≤ 2	≥ 8760
NARE	1973	Y + I	> 2 … ≤ 30	> 1 … < 8760
NEMO	2011	Y + I	≤ 2	≥ 8760
NEMS	1993	M	> 2 … ≤ 30	≤ 1
NEWAVE	1992	Y	> 2 … ≤ 30	> 1 … < 8760
ObjECTS-MiniCAM	2005	M	> 30	≥ 8760
Oemof (SOLPH)	2017	M	> 2 … ≤ 30	> 1 … < 8760
OpenDSS	2009	M	≤ 2	≥ 8760
ORCED	1996	Y + I	≤ 2	≥ 8760
OSeMOSYS	2011	I	> 30	> 1 … < 8760
OVER/UNDER	1977	I	> 2 … ≤ 30	≤ 1
PLEXOS	2003	Y + I	> 2 … ≤ 30	≥ 8760
POLES	1993	M	> 30	≤ 1
PowerGAMA	2013	M	≤ 2	≥ 8760
POWERS	2001	M	> 30	> 1 … < 8760
PRIMES	1994	M	> 30	≤ 1
ProdRisk	1994	Y + I	> 2 … ≤ 30	≥ 8760
PROMETHEUS	2007	M	> 30	≤ 1
PROVIEW	1980	Y + I	> 2 … ≤ 30	≥ 8760
PyPSA	2016	Y + I	≤ 2	≥ 8760
RAMSES	1988	Y	> 2 … ≤ 30	≥ 8760
RAPSim	2014	M	≤ 2	≥ 8760
Ready Reckoner	1997	M	> 2 … ≤ 30	> 1 … < 8760
ReEDS	2018	M	> 30	> 1 … < 8760
ReMIND	2015	M	> 30	> 1 … < 8760

(continued)

Table A.1 (continued)

Name of the modeling system	Year of release	Purpose	Time horizon, years	Number of intervals per year
REMARK	2006	Y	≤ 2	≥ 8760
REMix	2012	Y + I	≤ 2	≥ 8760
renpass	2015	Y	≤ 2	≥ 8760
RETScreen	1996	I	> 30	$> 1 \ldots < 8760$
ROM	2007	Y	≤ 2	≥ 8760
RSTEM	2005	Y	$> 2 \ldots \leq 30$	$> 1 \ldots < 8760$
SAFIRE	1997	I	> 30	≤ 1
SAGE	2002	I	$> 2 \ldots \leq 30$	≤ 1
SAM	2017	I	≤ 2	≥ 8760
SAMLAST	1993	Y	≤ 2	$> 1 \ldots < 8760$
SESAM	1987	M	$> 2 \ldots \leq 30$	$> 1 \ldots < 8760$
SIMPOW	2019	M	≤ 2	≥ 8760
SIREN	2014	M	≤ 2	≥ 8760
SIVAEL	1990	M	≤ 2	≥ 8760
SMART	2009	Y + I	> 30	≥ 8760
SNOW	2015	M	> 30	$> 1 \ldots < 8760$
SPSEK	1984	Y + I	$> 2 \ldots \leq 30$	≤ 1
stELMOD	2013	Y	≤ 2	≥ 8760
STREAM	2004	M	≤ 2	≥ 8760
System optimizer	2004	I	$> 2 \ldots \leq 30$	$> 1 \ldots < 8760$
SWITCH	2012	Y + I	$> 2 \ldots \leq 30$	≥ 8760
Temoa	2013	M	$> 2 \ldots \leq 30$	≤ 1
TRNSYS16	1975	Y + I	> 30	≥ 8760
TRNSYS18	2018	M	> 30	≥ 8760
UniSyD3.0	2003	M	> 30	$> 1 \ldots < 8760$
UPLAN	1988	Y	$> 2 \ldots \leq 30$	≥ 8760
Urbs	2019	Y + I	> 30	$> 1 \ldots < 8760$
UREM	2007	I	> 30	≤ 1
VTT-EMM	2007	Y	$> 2 \ldots \leq 30$	$> 1 \ldots < 8760$
WASP	1972	I	$> 2 \ldots \leq 30$	$> 1 \ldots < 8760$
WEM	1993	I	$> 2 \ldots \leq 30$	≤ 1
WeSIM	2013	Y + I	$> 2 \ldots \leq 30$	≥ 8760
WILMAR Planning Tool	2006	Y	≤ 2	≥ 8760
WITCH	2005	I	> 30	≤ 1

Table A.2 Systems for modeling economic equilibrium of energy complexes

AURORAxmp	LIBEMOD	PLEXOS
BALMOREL	MARKAL/TIMES	POLES
E4cast	NEMO	PRIMES
ETM model	NEMS	ReEDS
ENPEP	Oemof (SOLPH)	SNOW
ETSAP-TIAM	ObjCETS-MiniCFM	TIMES
GCAM	ORCED	UniSyD3.0
GEM-E3	PERSEUS	WITCH

Table A.3 Mathematical problem solvers

Modeling system	Solver	Developer	Year of solver development
NEMS	MPSIII optimization and modeling library	Ketron optimization	1980
PLEXOS	Xpress-MP	FICO	1988
	CPLEX	IBM	1988
	MOSEK	MOSEK	2000
	GUROBI	Gurobi	2009
PRIMES	GAMS	GAMS Dev. Cor. IBM	1987
	CPLEX		1988
	PATH	Wisconsin Uni	1995

Table A.4 Modern energy equilibrium modeling systems

Modeling system	Country of origin	Solver
IFFS	USA	PIES, Gauss–seidel algorithm
NEMS	USA	PIES, block gauss–seidel
ENPEP	USA	Data not found
GEMS	USA	Data not found
GEMINI	France	PATH
PRIMES	Greece	PATH
C-GEM	China, USA	PATH

Appendix B
Sets, Parameters, and Variables of the Model Developed for the Energy System of Ukraine

Sets

T modeling horizon in hours, weeks, years;
WPP list of hydroelectric power plant generators;
TCC set of thermal power plant units;
HP list of pumped storage hydroelectric generators.

Parameters

DATEmodeling date.

Parameters for Nuclear Power Plants (NPP), Solar Power Plants (SPP), and Wind Power Plants (WPP):

C^C_{NPP} Cost of supplying electricity to the grid by a nuclear power plant unit, in USD/MWh;
C^C_{PV} Cost of supplying electricity to the grid by a solar power plant unit, in USD/MWh;
C^C_{WIND} Cost of supplying electricity to the grid by a wind power plant unit, in USD/MWh.
P^{Σ}_{NPP} Total power supplied to the grid by all nuclear power plants, in MW;
$P^{\sum G}_{PV}$ Maximum allowable generation capacity of all solar power plants, in MW;
$P^{\sum G}_{WIND}$ Maximum allowable generation capacity of all wind power plants, in MW;
K^G_{PV} Forecasted growth factor for generation capacity of all solar power plants;
K^G_{WIND} Forecasted growth factor for generation capacity of all wind power plants.

A. Zaporozhets et al., *Structure Optimization of Power Systems with Renewable Energy Sources*, Studies in Systems, Decision and Control 583,
https://doi.org/10.1007/978-3-031-83697-8

Parameters for Hydropower Plants (WPP):

P_{WPP} Installed capacity of WPP units, in MW;
P_{WPP}^{MAX} Maximum power that an WPP unit supplies to the grid, in MW;
P_{WPP}^{MIN} Minimum power that an WPP unit supplies to the grid, in MW.
$P_{WPP}^{RUMP\_MAX}$ Maximum possible change in power of an WPP unit while operating, in MW;
$P_{WPP}^{\sum G}$ Maximum allowable level of power generation by all WPPs, in MW;
$E_{WPP}^{D\_MAX}$ Maximum amount of energy that an WPP supplies to the grid over the course of a day, in MWh;
$E_{WPP}^{D\_MIN}$ Minimum amount of energy that an WPP supplies to the grid over the course of a day, in MWh;
C_{WPP}^{G} Cost of supplying electricity to the grid by an WPP unit, in USD/MWh.

Parameters for Thermal Power Plants (TPP):

P_{TPP} Installed capacity of TPP units, in MW;
K_{TPP} Number of TPP units of each type;
P_{TPP}^{MAX} Maximum power that a TPP unit can deliver to the grid, in MW;
P_{TPP}^{MIN} Minimum power that a TPP unit can deliver to the grid, in MW;
$P_{TPP}^{RUMP\_MIN}$ Minimum possible change in power of a TPP unit while operating, in MW;
$P_{TPP}^{RUMP\_MAX}$ Maximum possible change in power of a TPP unit while operating, in MW;
C_{TPP}^{G} Cost of supplying electricity to the grid from a TPP unit, in USD/MWh;
k_{TPP}^{GD} Number of starts of TPP units per day;
$k_{TPP}^{G\_MIN}$ Minimum number of TPP units carrying load;
$k_{TPP}^{G\_MAX}$ Maximum number of TPP units carrying load;
τ_{TPP}^{GD} Duration (number of hours) of operation of a TPP unit per day.
R_{TPP}^{SWITCH} 1use reserve capacity with TPP, 0do not use;
R_{TPP}^{UP} 1consider the minimum reserve value for load with TPP, 0do not consider;
R_{TPP}^{DOWN} 1consider the minimum reserve value for unloading with TPP, 0do not consider;
$R_{TPP}^{MIN\_UP}$ Minimum reserve value for load with TPP, in MW;
$R_{TPP}^{MIN\_DOWN}$ Minimum reserve value for unloading with TPP, in MW.

Parameters for Pumped-Storage Hydroelectric Plants (HP):

P_{HP} Power of HP units in pumping mode, in MW;
G_{HP} Power of HP units in generating mode, in MW;
C_{HP}^{G} Cost of using HP units in generating mode, in USD/MWh;
C_{HP}^{P} Cost of using HP units in pumping mode, in USD/MWh.
V_{HP}^{P} Maximum amount of energy stored by HP units in pumping mode, in MWh;
τ_{HP}^{PG} Duration (number of hours) of continuous operation of HP units in both pumping and generating modes;

k_{HP}^{DPG} Number of starts of HP units per day in both pumping and generating modes;

O_{HP}^{D} Restriction on the number of starts per day for HP units: 1—each unit must be started once per day, 0—the condition is not required.

V_{HP}^{WD} Balance of water accumulation and use throughout the day for HP units: 1—each unit must use all accumulated water, 0—the condition is not required;

$P_{HP}^{\sum P}$ Maximum allowable power consumption for pumping by all HPs, in MW;

$P_{HP}^{\sum G}$ Maximum allowable power generation by all HPs, in MW;

O_{HP}^{ST} Restriction on simultaneous operation of HP units: 1—pumping and generating cannot occur simultaneously; if a unit is pumping at a given time, all other units must not generate, 0—the condition is not required.

R_{HP}^{SWITCH} 1—Use reserve capacity with HP, 0—Do not use;

R_{HP}^{UP} 1—Consider the maximum reserve capacity for load with HP, 0—Do not consider;

R_{HP}^{DOWN} 1Consider the minimum reserve capacity for unloading with HP, 0Do not consider;

$R_{HP}^{MAX\_UP}$ Maximum reserve capacity for load with HP, in MW;

$R_{HP}^{MIN\_DOWN}$ Minimum reserve capacity for unloading with HP, in MW.

Parameters for Energy Storage Systems (Batteries):

V_B Installed capacity—the total capacity of the storage system, in MWh;

$P_B^{G\_MAX}$ Maximum discharge power—the maximum power that the storage system can deliver to the grid, in MW;

$P_B^{C\_MAX}$ Maximum charge power—the maximum power that the storage system can draw from the grid during charging, in MW;

$E_B^{D\_MAX}$ Maximum allowable charge–discharge volume per day—the maximum amount of energy that can be charged and discharged by the storage system in one day, in MWh;

C_B^G Cost of delivering electricity to the grid—the cost per MWh of supplying energy to the grid from the storage system, in USD/MWh;

C_B^C Cost of consuming electricity from the grid—the cost per MWh of charging the storage system from the grid, in USD/MWh.

R_B^{SWITCH} 1—Use storage for reserve: 1—Use storage for reserve, 0—Do not use;

R_B^{UP} 1—Consider maximum reserve for load using storage: 1—Consider the maximum reserve value for load using storage, 0—Do not consider;

R_B^{DOWN} 1—Consider maximum reserve for unload using storage: 1Consider the maximum reserve value for unload using storage, 0—Do not consider;

$R_B^{MAX\_UP}$ Maximum reserve value for load using storage: Maximum reserve capacity for load, in MW;

$R_B^{MAX\_DOWN}$ Maximum reserve value for unload using storage: Maximum reserve capacity for unload, in MW;

E_B^τ Charge–discharge balance of the storage system over previous hours: 1—The total charge of the battery should be greater than the total discharge over the previous hours, 0—The total charge of the battery should be greater than zero.

Parameters for General Reserve:

R_{ALL}^{UP} 1—Consider minimum general reserve for load: 1—Consider the minimum general reserve value for load, 0—Do not consider;

R_{ALL}^{DOWN} 1—Consider minimum general reserve for unload: 1—Consider the minimum general reserve value for unload, 0—Do not consider;

$R_{ALL}^{MIN\_UP}$ Minimum general reserve value for load: Minimum reserve capacity for load, in MW;

$R_{ALL}^{MIN\_DOWN}$ Minimum general reserve value for unload: Minimum reserve capacity for unload, in MW.

Parameters for Import-Export:

Minimum export capacity: Minimum power for export, in MW.

P_I^{MAX} Maximum import power, in MW;
P_E^{MAX} Maximum export power, in MW;
P_I^{MIN} Minimum import power, in MW;
P_E^{MIN} Minimum export power, in MW.
$E_I^{D\_MAX}$ Maximum permissible import volume per day, in MWh;
$E_E^{D\_MAX}$ Maximum permissible export volume per day, in MWh;
$E_I^{D\_MIN}$ Minimum permissible import volume per day, in MWh;
$E_E^{D\_MIN}$ Minimum permissible export volume per day, in MWh.
C_I^G Cost of import, in USD/MWh;
C_E^G Cost of export, in USD/MWh;
E_{IE}^D Balance of import–export volumes: 1—zero balance of flows (import equals export), 0—zero balance condition is not enforced.

Penalty Function Parameters:

$C_{deficit}^P$ Cost of penalty for insufficient power coverage;
$C_{overplus}^P$ Cost of penalty for excess power coverage.

Variables

Variables of the optimization model used for optimizing the hourly operation of the energy system to cover the load schedule.

Variables used for optimizing the hourly operation of NPPs, WPPs, SPPs, and WindPPs:

P_t^{NPP} total hourly power output from all NPPs to the grid, MW;

P_t^{WPP} total hourly power output from all WPPs to the grid, MW;
P_t^{PV} total hourly power output from all SPPs to the grid, MW;
P_t^{WIND} total hourly power output from all WPPs to the grid, MW.

Variables used in the optimization of hourly operation of TPPs:

P_t^{TPP} hourly power output that each TPP unit delivers to the grid, MW;
Y_{bt}^{TPP} start indicator (activation) of a TPP unit or multiple units under load, binary variable, 1—start, the sum of these variables indicates the number of starts of a unit per day;
Y_{et}^{TPP} stop indicator (deactivation) of a TPP unit or multiple units under load, binary variable, 1—stop;
Y_{ot}^{TPP} number of TPP units covering the load schedule in a specific time period after starting in the previous period;
Y_{Lt}^{TPP} indicator of participation of a TPP unit in covering the load schedule at a specific time.

Variables used in the hourly operation optimization of HPs:

P_t^{HP} hourly power consumed from the grid by each HP unit in pumping mode, MW;
G_t^{HP} hourly power supplied to the grid by each HP unit in generation mode, MW;
$Y_{bt}^{HP\_P}$ indicator of the start (activation) of a HP unit in pumping mode, binary variable;
$Y_{bt}^{HP\_G}$ indicator of the start (activation) of a HP unit in generation mode, binary variable;
E_t^{HP} current amount of energy stored by each HP, MWh;
$E_t^{HP\_\Sigma}$ total current amount of energy stored by all HPs, MWh.

Variables used in the hourly operation optimization of the battery storage:

P_t^{BC} hourly power consumed from the grid by the storage unit during charging, MW;
P_t^{BG} hourly power supplied to the grid by the storage unit during discharging, MW;
E_t^{BC} E_t^{BG}current amount of energy stored by the storage unit during charging and discharging, MWh.

Variables for hourly use of import and export:

P_t^I hourly import power into the grid, MW;
P_t^E hourly export power from the grid, MW.

Variables for hourly reserve for system load and unload:

$R_t^{TPP\_UP}$ Hourly reserve power of TPPs for load, MW;
$R_t^{TPP\_DOWN}$ Hourly reserve power of TPPs for unload, MW;
$R_t^{HP\_UP}$ Hourly reserve power of HPs for load, MW;
$R_t^{HP\_DOWN}$ Hourly reserve power of PSHPs for unload, MW;
$R_t^{B\_UP}$ $R_t^{B_DOWN}$Hourly reserve power of the battery for load, unload, MW;
$R_t^{\Sigma\_UP}$ $R_t^{\Sigma_DOWN}$Hourly total system reserve power for load, unload, MW.

Penalty variables for insufficient or excessive hourly system power:

$P_t^{P\_deficit}$ Amount of insufficient coverage power, MW;
$P_t^{P\_overplus}$ Amount of excessive coverage power, MW.

Variable of required hourly power to cover the electrical load schedule of the system:

D_t Required hourly power (MW) to cover the system load schedule.

Appendix C
Parameters of Electrical Power Systems of Ukraine and ENTSO-E Member Countries for 2018

See Tables C.1 and C.2.

A. Zaporozhets et al., *Structure Optimization of Power Systems with Renewable Energy Sources*, Studies in Systems, Decision and Control 583,
https://doi.org/10.1007/978-3-031-83697-8

Table C.1 Electrical power (MW) for 2018, showing the peaks and troughs of load for the power systems of Ukraine and neighboring ENTSO-E member countries

	Peaks	Troughs						
Hours	Poland 28–02	Ukraine 20–12	Romania 27–02	Hungary 02–03	Poland 20–05	Ukraine 27–05	Romania 08–04	Hungary 21–05
1	20,025	18,990	7316	5347	14,271	13,475	5320	3437
2	19,365	19,013	7247	5112	13,588	12,897	5115	3286
3	19,063	19,258	7232	4891	13,183	12,429	5039	3150
4	19,106	20,031	7309	4759	13,031	12,172	4988	3089
5	19,474	21,602	7472	4964	12,551	12,067	4941	3025
6	20,350	22,760	7946	5421	12,262	12,282	4877	3015
7	22,455	23,916	8684	6054	12,757	12,794	4863	3132
8	24,431	24,422	9259	6318	13,501	12,879	4986	3089
9	25,582	24,632	9529	6395	14,591	13,459	5018	3591
10	25,806	24,375	9596	6479	15,379	14,025	4964	3840
11	25,604	24,180	9525	6541	15,690	13,937	4859	3927
12	25,759	24,306	9453	6592	15,698	14,122	4760	3986
13	25,766	23,336	9376	6587	15,640	14,064	4650	3965
14	25,700	24,499	9128	6531	15,538	14,103	4542	3865
15	25,387	24,684	8940	6404	15,327	14,022	4548	3911
16	24,903	24,357	8962	6368	14,930	13,858	4553	3904
17	24,737	23,909	9021	6305	14,762	14,073	4642	3950
18	25,435	23,579	9526	6395	14,749	14,087	4764	3980
19	26,258	23,132	9695	6410	15,104	14,095	5033	4040
20	26,081	22,703	9469	6358	15,473	13,951	5748	4111
21	25,439	21,385	9050	6149	15,941	14,453	5881	4245
22	23,893	20,719	8338	5979	16,567	15,173	5579	4230
23	22,307	19,895	7779	5710	15,940	14,273	5251	3935
24	20,934	19,308	7457	5523	14,847	13,395	4980	3693

Table C.2 Parameters of participation in covering the total minimum and maximum load of the ENTSO-E energy system on June 17 and February 28, 2018, for the energy systems of Ukraine and neighboring ENTSO-E member countries

	Participation in covering the total maximum load (MW), 28.02.2018					Participation in covering the total minimum load (MW), 17.06.2018				
Hours	Poland	Ukraine	Romania	Hungary	Slovakia	Poland	Ukraine	Romania	Hungary	Slovakia
1	20,025	19,112	7271	5333	3646	15,072	14,694	5583	4070	2681
2	19,365	18,907	7183	5049	3551	14,316	14,820	5423	3906	2576
3	19,063	18,884	7164	4840	3527	13,987	14,724	5273	3727	2542
4	19,106	19,329	7273	4731	3523	13,504	14,658	5263	3650	2531
5	19,474	20,054	7457	4959	3637	13,051	14,865	5122	3588	2514
6	20,350	21,707	7969	5410	3787	12,896	15,112	5086	3500	2496
7	22,455	22,611	8661	6184	4167	13,397	15,012	5237	3613	2601
8	24,431	23,791	9273	6262	4336	14,302	14,228	5594	3905	2776
9	25,582	24,237	9511	6321	4409	15,367	14,960	5782	4175	2986
10	25,806	24,139	9509	6326	4432	16,269	15,585	5909	4409	3170
11	25,604	23,906	9366	6323	4332	16,786	15,816	5938	4590	3285
12	25,759	23,734	9453	6350	4356	16,967	16,200	6018	4693	3332
13	25,766	23,587	9297	6322	4304	16,983	16,158	6059	4733	3214
14	25,700	23,467	9011	6296	4276	16,856	16,157	5960	4627	3134
15	25,387	23,521	8842	6234	4248	16,431	16,230	5943	4589	3072
16	24,903	23,189	8820	6228	4196	16,236	16,228	6004	4589	3039
17	24,737	23,896	8894	6207	4241	16,070	16,210	6084	4619	3040
18	25,435	24,111	9385	6448	4323	16,094	16,312	6073	4624	3032
19	26,258	23,782	9606	6563	4413	16,151	16,302	6125	4605	3013
20	26,081	23,295	9449	6537	4381	16,570	16,091	6202	4622	3058

(continued)

Table C.2 (continued)

	Participation in covering the total maximum load (MW), 28.02.2018					Participation in covering the total minimum load (MW), 17.06.2018				
Hours	Poland	Ukraine	Romania	Hungary	Slovakia	Poland	Ukraine	Romania	Hungary	Slovakia
21	25,439	22,440	9037	6374	4291	16,777	16,380	6492	4675	3090
22	23,893	21,324	8389	6183	4053	17,210	16,804	6341	4904	3130
23	22,307	20,150	7823	5875	3843	17,057	15,678	5856	4717	2993
24	20,934	19,492	7482	5686	3705	15,897	15,360	5464	4392	2848

The manufacturer's authorised representative in the EU is Springer
Nature Customer Service Centre GmbH, Europaplatz 3, 69115 Heidelberg,
Germany. If you have any concerns regarding our products, please
contact ProductSafety@springernature.com

Printed and bound by CPI Group (UK) Ltd, Croydon, CR0 4YY
01/10/2026
02225977-0011